T0399831

# MANUFACTURING AND INDUSTRIAL ENGINEERING

# Science, Technology, and Management Series

*Series Editor: J. Paulo Davim, Professor, Department of Mechanical Engineering, University of Aveiro, Portugal*

This book series focuses on special volumes from conferences, workshops, and symposiums, as well as volumes on topics of current interested in all aspects of science, technology, and management. The series will discuss topics such as, mathematics, chemistry, physics, materials science, nanosciences, sustainability science, computational sciences, mechanical engineering, industrial engineering, manufacturing engineering, mechatronics engineering, electrical engineering, systems engineering, biomedical engineering, management sciences, economical science, human resource management, social sciences, engineering education, etc. The books will present principles, models techniques, methodologies, and applications of science, technology and management.

**Optimization Using Evolutionary Algorithms and Metaheuristics**
*Edited by Kaushik Kumar and J. Paulo Davim*

**Integration of Process Planning and Scheduling**
Approaches and Algorithms
*Edited by Rakesh Kumar Phanden, Ajai Jain and J. Paulo Davim*

**Understanding CATIA**
A Tutorial Approach
*Edited by Kaushik Kumar, Chikesh Ranjan, and J. Paulo Davim*

**Manufacturing and Industrial Engineering**
Theoretical and Advanced Technologies
*Edited by Pakaj Agarwal, Lokesh Bajpai, Chandra Pal Singh, Kapil Gupta, and J. Paulo Davim*

**Multi-Criteria Decision Modelling**
Applicational Techniques and Case Studies
*Edited by Rahul Sindhwani, Punj Lata Singh, Bhawna Kumar, Varinder Kumar Mittal, and J. Paulo Davim*

**High-k Materials in Multi-Gate FET Devices**
*Edited by Shubham Tayal, Parveen Singla, and J. Paulo Davim*

**Advanced Materials and Manufacturing Processes**
*Edited by Amar Patnaik, Malay Kumar, Ernst Kozeschnik, Albano Cavaleiro, J. Paulo Davim, and Vikas Kukshal*

**Computational Technologies in Materials Science**
*Edited by Shubham Tayal, Parveen Singla, Ashutosh Nandi, & J. Paulo Davim*

For more information about this series, please visit: https://www.routledge.com/Science-Technology-and-Management/book-series/CRCSCITECMAN

# MANUFACTURING AND INDUSTRIAL ENGINEERING
## THEORETICAL AND ADVANCED TECHNOLOGIES

Edited by
Pankaj Agarwal
Lokesh Bajpai
Chandra Pal Singh
Kapil Gupta
J. Paulo Davim

CRC Press
Taylor & Francis Group
Boca Raton London New York

CRC Press is an imprint of the
Taylor & Francis Group, an **informa** business

First edition published 2021
by CRC Press
6000 Broken Sound Parkway NW, Suite 300, Boca Raton, FL 33487-2742

and by CRC Press
2 Park Square, Milton Park, Abingdon, Oxon, OX14 4RN

© 2021 selection and editorial matter, Pankaj Agarwal, Lokesh Bajpai, Chandra Pal Singh, Kapil Gupta, and J. Paulo Davim; individual chapters, the contributors

CRC Press is an imprint of Taylor & Francis Group, LLC

Reasonable efforts have been made to publish reliable data and information, but the author and publisher cannot assume responsibility for the validity of all materials or the consequences of their use. The authors and publishers have attempted to trace the copyright holders of all material reproduced in this publication and apologize to copyright holders if permission to publish in this form has not been obtained. If any copyright material has not been acknowledged please write and let us know so we may rectify in any future reprint.

Except as permitted under U.S. Copyright Law, no part of this book may be reprinted, reproduced, transmitted, or utilized in any form by any electronic, mechanical, or other means, now known or hereafter invented, including photocopying, microfilming, and recording, or in any information storage or retrieval system, without written permission from the publishers.

For permission to photocopy or use material electronically from this work, access www.copyright.com or contact the Copyright Clearance Center, Inc. (CCC), 222 Rosewood Drive, Danvers, MA 01923, 978-750-8400. For works that are not available on CCC please contact mpkbookspermissions@tandf.co.uk

*Trademark notice*: Product or corporate names may be trademarks or registered trademarks and are used only for identification and explanation without intent to infringe.

ISBN: 978-0-367-54174-3 (hbk)
ISBN: 978-0-367-54175-0 (pbk)
ISBN: 978-1-003-08807-3 (ebk)

Typeset in Times
by MPS Limited, Dehradun

# Contents

# Preface

Manufacturing and Industrial Engineering have been the two most significant streams of engineering that underpinned industrial revolutions and largely contributed towards economic growth. Research, developments and innovations are the essential activities that lay the foundation of technological evolution. In both manufacturing and industrial engineering, conventional practices, processes and techniques are currently being replaced by their modern and advanced versions in order to meet the productivity, safety, quality and sustainability requirements.

This book provides a basic theoretical understanding and discussion of technological advancements in some of the vital topics of manufacturing and industrial engineering. It comprehensively covers working principles, technical details, process mechanisms, reviews of literature, salient features, case studies, research results, current advancements and future directions.

This book consists of a total of 16 selected chapters on theoretical and advanced technologies in manufacturing and industrial engineering. It starts with Chapter 1 by providing an overview of the role of Cloud Manufacturing in modern industries. Chapter 2 sheds light on the modern combination of composite materials using bonding and cladding. Chapter 3 discusses green manufacturing basics and recent developments. Chapter 4 presents a systematic review of various additive manufacturing techniques. Post-processing of engineered parts and components made by additive manufacturing is detailed in Chapter 5. Chapter 6 reports the fabrication of composites by stir casting method along with a further study on its microstructure and mechanical and corrosion behaviour. Intelligent fault diagnosis of roller bearings is examined in Chapter 7. Finite element analysis based-sheet metal quality improvement is detailed in Chapter 8. Chapter 9 provides an analysis of human bone and biomaterials for bone implantation. The implementation and effectiveness of hybrid optimisation in case of supply chain management is investigated in Chapter 10. Advanced design and manufacturing of car door panels are examined in Chapter 11. Chapter 12 considers the manufacturing of modern composites using banana fibre along with the latest research in this area. Chapter 13 focuses on post-processing and corrosion behaviours of parts made by metal additive manufacturing. Nonlinear control strategies in electrohydraulic actuator systems and relevant past work are further discussed in Chapter 14. Chapter 15 reports the results of experimental work conducted on the evaluation of mechanical properties of photopolymer structures fabricated by scan-based microstereolithography. Finally, Chapter 16 concentrates on the various aspects of computer integrated manufacturing.

We hope that this book will be beneficial to researchers, scholars, professors, technical experts and specialists working in the area of mechanical, manufacturing and industrial engineering. We sincerely acknowledge CRC Press for this opportunity and their professional support. Finally, we would like to thank all contributors for their time and efforts.

Pankaj Agarwal, Lokesh Bajpai,
Chandra Pal Singh, Kapil Gupta, J. Paulo Davim

# Editor Biographies

**Pankaj Agarwal** is working as Professor in the Mechanical Engineering Department at Samrat Ashok Technological Institute Vidisha (MP) India. He obtained a PhD in Mechanical Engineering from Barkhatullah University, Bhopal, India. Currently, he is the Head of Mechanical Engineering Department. Additive Manufacturing, 3D printing, Industry 4.0, micromanufacturing, Lean manufacturing and flexible manufacturing system, are his pertinent fields of interest. He has guided several ME and PhD theses in Advanced Production System and Computer Integrated Manufacturing. He has authored more than 100 articles in journals and international conferences. He bears experience of more than 28 years in research, teaching and training in the field of mechanical and industrial engineering. He also has an interest in AI in Manufacturing, IOT in Manufacturing, Digital Twins and ERP solutions.

    **Lokesh Bajpai** is working as Professor in the Mechanical Engineering Department of the Samrat Ashok Technological Institute Vidisha (MP) India. He has 36 years of teaching and research experience. He has guided several ME and PhD theses in manufacturing area, particularly in advanced production and computer integrated manufacturing. He obtained his PhD from Barkhatullah University, Bhopal India. Advanced machining processes, flexible manufacturing system, supply chain management and 3D printing are his specialisations. He has authored a significant amount of articles in journals and conferences. He has been conducting industrial research projects and training programs in various topics, i.e. TQM, supply chain management, production planning and control and maintenance management of industrial engineering.

    **J. Paulo Davim** received his PhD in Mechanical Engineering in 1997, MSc degree in Mechanical Engineering with a focus on Materials and Manufacturing Processes in 1991, Mechanical Engineering degree (five years) in 1986 from the University of Porto (FEUP), Aggregate title (Full Habilitation) from the University of Coimbra in 2005 and DSc from the London Metropolitan University in 2013. He was conferred the title Eur Ing by FEANI-Brussels and Senior Chartered Engineer by the Portuguese Institution of Engineers with an MBA and Specialist title in Engineering and Industrial Management. Currently, he is a Professor at the Department of Mechanical Engineering of the University of Aveiro, Portugal. He has more than 30 years of teaching and research experience in manufacturing, materials, mechanical and industrial engineering, with a special emphasis in machining and tribology. He also has an interest in management, engineering education and higher education for sustainability.

    **Kapil Gupta** is working as an Associate Professor in the Department of Mechanical and Industrial Engineering Technology at the University of Johannesburg. He obtained his PhD in Mechanical Engineering with a specialisation in Advanced Manufacturing. Advanced machining processes, sustainable manufacturing, green machining, precision engineering and gear technology are his areas of expertise. He has authored several SCI/ISI Journal and International Conference articles. He has also authored and edited 12 international books on hybrid machining, advanced gear manufacturing, micro and precision manufacturing and sustainable manufacturing. He is an editorial board member of a significant amount of international journals. He has also delivered keynote lectures, distinguished speeches in international conferences and symposiums and seminar talks at international universities.

**Chandra Pal Singh** is working as an Assistant Professor in the Mechanical Engineering Department in Samrat Ashok Technological Institute Vidisha (MP) India. He has 18 years of teaching experience. He acquired a PhD in Mechanical Engineering from Maulana Azad National Institute of Technology, Bhopal, India. Metal forming, deep drawing, fatigue failure and finite element analysis are his areas of focus. He has authored several articles published in international journals and conference proceedings. He is currently supervising many PG students who are conducting cutting-edge research in the advanced manufacturing field.

# 1 An Overview on the Role of Cloud Manufacturing in Modern Industries

*V. Kavimani and P. M. Gopal*

## CONTENTS

## 1.1  INTRODUCTION

Cloud Manufacturing (CM) is emerging as a newer ethos in the manufacturing-based industrial sector. A comprehensive concept of CM was first presented by Li, Zhang, and Chai (2010). The most outstanding and favourable feature of CM is unified and suitable allotment of various classes of dispersed developing possessions, spearheading the concept of manufacturing as a service (Li, Zhang, and Chai 2010; Wu and Yang 2010; Zhang et al. 2011). Cloud manufacturing aids suppliers in efficiently summarising and categorising manufacturing assets and competencies in making these accessible as facilities for clients in a workers-running-based manufacturing cloud system.

The concept is based on the premise that enterprises might acquire several manufacturing facilities with the support of the internet, comparable to attaining electricity and water in our everyday life. The CM process contains highly assorted manufacturing assets that can be collected thru various clients, minor jobs and also for difficult cooperative manufacturing missions (Zhang et al. 2010; Xu, Wang, and Newman 2011). The capabilities of CM are geared towards assisting in the service station of the entire production of product improvement life sequence, and it covers an extensive range of cloud facilities, from the exploration of customer requirements and simulation, resource scheduling, supply-chain control and product design manufacturing, generally, the processes that manage the components of failure (Tao et al. 2011; Zhang et al. 2014). A comprehensive Cloud Manufacturing System (CMS) is comprised of three groups – shareholders (specifically, workers), suppliers and customers. It is through their teamwork that the sustainable task of CMS is conserved (Figure 1.1).

This chapter deals with the insights on the role of cloud manufacturing in different industrial processes and scenarios as follows:

* Role of CM in the production scheduling system
* Role of CM in product customisation
* Role of CM in product lifecycle management
* Role of CM in rapid product development by additive manufacturing

## 1.2  THE ROLE OF CLOUD MANUFACTURING IN PRODUCTION SCHEDULING SYSTEM

### 1.2.1  THE PROCEDURE OF SCHEDULING IN CLOUD MANUFACTURING

The scheduling process in CMS consists of five major moduli namely, job submission, initial job processing, scheduling, outcome dispatch and service valuation (Figure 1.2).

#### 1.2.1.1  Job Submission

The entire scheduling process initiates the orders or jobs placed by the customer. Based on purpose and necessity, jobs can be categorised into design jobs, testing jobs, manufacturing jobs, etc.

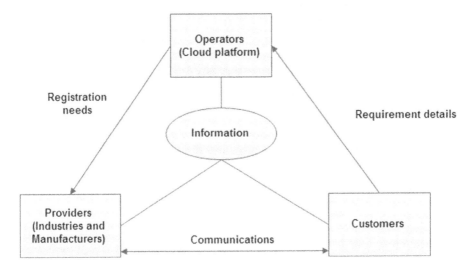

**FIGURE 1.1** Conceptual Models for Cloud Manufacturing.

### 1.2.1.2 Initial Job Processing

After customers submit jobs in the cloud platform, an initial procedure that includes depiction, primarily classification, disintegration, analysis, etc. begins. Once initial processing is complete, serviceable and non-serviceable necessities are clearly included in each job requirement. The utility forms referred are essential to be able to understand the aims of finishing jobs. Recognizing this purpose involves supplication and implementation of essential categories of services (Mourtzis et al. 2016).

### 1.2.1.3 Scheduling

After initial job processing, the schedule management module will proceed to convey job timelines for the scheduling support section, the service management section and the monitor controlling section. The core development modules are accountable for creating optimal plans and supervising job implementation

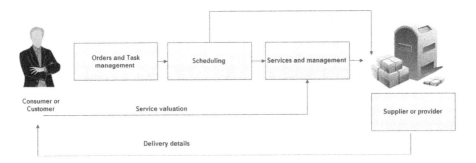

**FIGURE 1.2** The Steps Involved in Cloud Manufacturing Assisted Scheduling.

methods. The planning support section is mainly liable for maintaining the planning metrics, regulations, approaches, processes and delivery provisions for the planning management section.

The service managing section is responsible for accomplishing service-associated events. These events are essential for scheduling service categorisation, identical and exploration, configuration and valuation, etc. The observing management module studies the position of assets and jobs in workshop platforms and provides real-time information of resources and jobs that are essential for attaining improved scheduling. The procedures for scheduling implementation are described below.

Initially, optimal plans are created by systematic procedures; the scheduled jobs would be delivered to various suppliers. As implementation progresses, the source and status of the jobs are supervised by the assistance of real-time monitoring. Intermittent switches towards the assets of the CMS base are desired. Throughout the scheduling process, the company undertakes various sub-jobs of a task interrelates and interconnects with everyone in order to confirm the smoother implementation of the jobs (Liu et al. 2017).

### 1.2.1.4   Result Delivery

After accomplishing a job, related assets are dispatched, and the eventual implementation outcomes are conveyed to customers through the internet or by logistics.

### 1.2.1.5   Service Valuation

Subsequently, customers collect the finished outcomes, and they are provided with a chance to evaluate the facilities that they utilised. The assessment results show their general satisfaction rate with the results.

### 1.2.2   VARIOUS FEATURES OF PLANNING IN CM

CM is a material production hypothesis, and its progress varies in forms of manufacturing simulations, namely, network and agile-based manufacturing that are connected based on operative models and integrated technologies. Scheduling in cloud manufacturing has various features that are categorised into operation mode, resource and requirement and knowledge-related features (Mourad et al. 2020; Laili, Lin, and Tang 2020).

### 1.2.2.1   Operation Mode-Based Features

There are various classifications of shareholders in a CMS, and this mainly includes the suppliers, workers and customers, and a specific type of shareholder comprises many entities. The overview of workers for the integrated organisation and process of a CMS, separates the CM from earlier available manufacturing prototypes. Likewise, in relation to earlier manufacturing prototypes, more distant entities are included in a CMS rather than those implicated in earlier manufacturing models. Persons in each group of shareholders are independent individuals who determine and make their own choices (Wang, Zhang, and Qi 2017). Every individual has their own purpose and predilections; thus, planning intentions are further varied to minor-scale component production systems. Association, comprising an association of various suppliers and relationships among planning of cloud-based jobs and resident jobs also significant features for scheduling in CM (Zhou, Zhang, and Fang 2020).

### 1.2.2.2 Resource Requirement-Based Features

Extensive manufacturing resources on various suppliers are combined together in cloud basement, and CM summarises these for cloud facilities. Assessment with individuals in earlier material production systems constitutes a considerably higher range of possessions that are included in CM. The planning of higher range possessions requires additional effective scheduling approaches and procedures (Dong et al. 2020; Lin and Chong 2017). At the time when planning progresses, an enormous amount of resource data are amassed, and the gathering and analysis of these files might efficiently enhance proficiency in planning.

### 1.2.2.3 Knowledge Technology-Based Features

In CM, entire manufacturing resources remain condensed into cloud facilities with the support of technologies of servitisation and virtualization that can efficiently protect against the deviation of manufacturing assets and unanticipated problems in geographic deliveries of material reserves. Conversion of manufactured possessions into cloud-based facilities essentially distinguishes CM as diverse from earlier production prototypes (Y. Hu et al. 2020; Yang et al. 2020). Planning in CM is constructed on cloud facilities that could be expediently scored and organised in order to adjust these for prerequisite jobs thru various regularities.

### 1.2.3 Cloud-Assisted Scheduling System

Model-driven related architecture is an application-based software engineering methodology that is used to develop CAS systems. It consists of four classifications, namely, the computation and basement self-determining models, and the platform and implementation explicit model. The purposes of these stages are quite similar to the customer need analysis, system-based necessities and the execution stages of the system as illustrated in Figure 1.3a. Interactions and relationships that occur among the CAS system are illustrated in Figure 1.3b, and these consist of sales managers, shop floor managers and production planners. This may be achieved through job orders, track order development positioning and work floor manufacture details, and an enactment might be followed to be able to estimate actual time duration [20, 21].

## 1.3 ROLE OF CLOUD MANUFACTURING IN PRODUCT CUSTOMISATION

The development in communication and information technologies have conveyed considerable types of consumables and industrial components, further changing into multifaceted, networked and intelligent systems. Smart robots and smartphones are

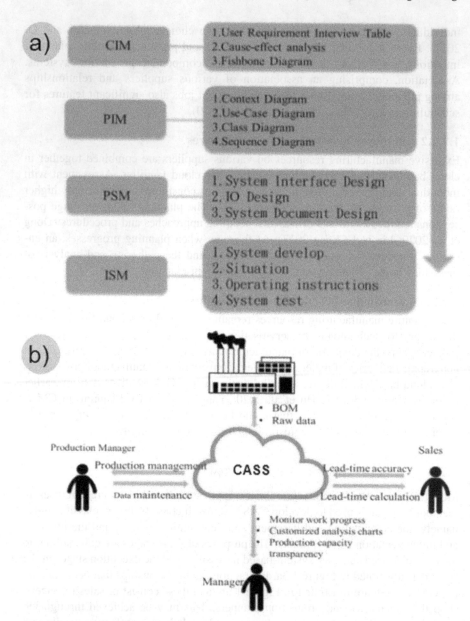

**FIGURE 1.3** Layout of a) model-driven architecture b) CAS system reprinted from Procedia Manufacturing, Vol. 17, Tzu-Han Hsu, Li-Chih Wang, Pei-Chun Chu, Development of a Cloud-Based Advanced Planning and Scheduling System, Pages No. 431, Copyright (2018), with permission from Elsevier.

an appropriate example of a smart product among industries and consumers alike. Additionally, service facilities are also a significant protagonist in the industrial sector. Such advancements expedite service because components are incorporated within a common product facility domain, thereby suggesting complete solutions for customers (Mont 2002; Porter and Heppelmann 2014). Apart from propensity towards efficiency, an alternative option in professional manufacture is customised. Further, the clients require additional individualized choices for preferred components. On the other hand, product developers also need to propose exclusively custom-made components to their clients in order to achieve competent authority in international level marketing and productivity improvement (Yang et al. 2017).

In order to attain this goal in a minimal time interval, product-based configuration systems are globally implemented in the customisation of products. Conversely, smart, connected products are classically integrated with smart assisted environments and interrelate only with other associated smart components based on connectivity by IOT, a factor made challenging for the manufacturers in customising smart products (Simeone, Deng, and Caggiano 2020; Wan et al. 2020). Additionally, the presence of entrenched service and functionalities will produce complications in the customisation of process interaction. In order to manage these types of concerns, smart products are embedded within service. This approach is referred to as Smart Service Product Customisation.

In the present situation, Smart Service Product Customisation based on smart facility appearance link-up is effective in gaining complete understanding of individual needs. In this chapter, a detailed survey revealed that the Smart Service Product Customisation approach, based on the perception of CM, could deliver consistent production ability and corresponding improved smart service (X. Li, Fang, and Yin 2020).

### 1.3.1 CLOUD-ASSISTED METHOD FOR SMART PRODUCT CUSTOMISATION

To keep up with distinct consumer needs, the customisation of smart service products is an essential factor that includes personalization of component-based, smart facility environments. In view of product outline, the prevailing investigation on product configuration systems primarily concentrates on product-physical-based functionalities without considering the smart environment and service associated features. Thus, highlighting the features of Smart Service Product Customisation, the functionality of product along with the organization of smartness (efficiency) and facilities are considered as illustrated in Figure 1.4.

The customisation process of products consists of three types of stakeholders. Customers a group of persons who require customised products and could be categorised into standard users with slight knowledge in design to skilled users with better than average design knowledge. In this case, the standard users might choose only the available configuration from the product configuration system, and skilful users have the capability to describe their own preferences in customisation beyond the product configuration system (Ma et al. 2020). Manufacturing initiatives have a major responsibility in providing material production possessions and competencies in understanding the fabrication of physical products that were formerly

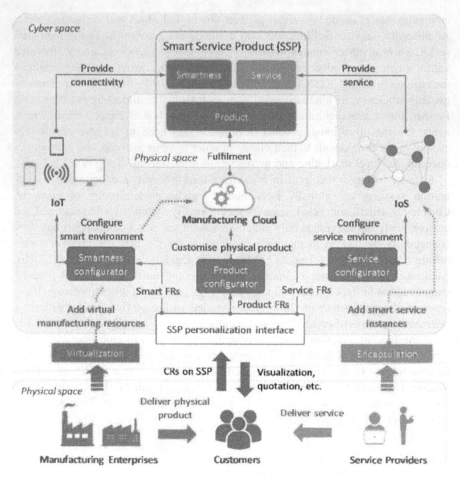

**FIGURE 1.4**   An Overview of Smart Product Customisation. Reprinted from Procedia CIRP, Vol. 72, Xinjuan Jin, Shiqiang Yu, Pai Zheng, Quan Liu, Xun Xu, Cloud-Based Approach for Smart ProductPersonalisation, Page No. 922–927, Copyright (2018), with permission from Elsevier.

assembled and distributed to consumers. Smart facility suppliers comprise of mediators who are in charge of smart facility-based usage with regards to smart service product customisation that improve product attractiveness. From the perception of cyberspace, as demonstrated in the framework, the manufacturing cloud, internet of things and internet of service offer technical support and infrastructure for the customisation of smart service products (Zhu et al. 2020).

A manufacturing cloud consists of a large number of manufacturing resources that are virtualized and interlinked with cloud facilities in order to allow the required provision for amenities. The virtualization of manufacturing resources conceives effective classification of the available data into computer-generated units to be able to categorize information into sensible units to accommodate differentiated needs, whereas the purpose of these logic units could be summarised with the help of service-oriented manufacturing concepts in assisting geographically distributed customers and recognising the sharing of manufacturing resources (Xia et al. 2020). Henceforth, abundant manufacturing cloud services are simple and could be easily reconfigurable in order to achieve modifications in manufacturing-based tasks, which will significantly upsurge independence of customisation in products.

Smart connectivity is mainly facilitated by the of internet of things (IOT). Herein, smart connections are classified into two different categories based on IOT, the assembly of smart products being the first category. Nominal cost, radio modules and compact storage devices made it conceivable to entrench digital memory (Pei Breivold 2020). In this connection, the smart environment helps in using the salient features of smart device that provides smart service product customizer. They are capable of identifying and regulating their pertinent situation by arresting and inferring optimal conditions and user actions, examining their choices, thus forestalling customer purpose and ultimately accomplishing target activities. Taking radio-frequency identification-based indoor technology as the example, the smartphone has the proficiency to identify where the customer is and can then respond consequently (Li et al. 2020; K. Wang 2020). If the customer is attending business meetings or programmes, then the customisable feature of smart mobiles automatically activates on the "DO NOT DISTURB" mode. Based on customisation procedure as depicted in Figure 1.4 through smart service product customisation interface, the user can order their necessities for personalising the smart service product that would be transformed into utility supplies in the form of product functionality, smart environment and on-demand service. Afterwards, three configurators, namely product configurator, service configurator and smartness configurator, are involved in designing and assembling the component of smart products based on the corresponding service environment (R. Hu et al. 2020).

## 1.3.2 INTERNET OF THINGS-BASED CLOUD MANUFACTURING APPROACH FOR PRODUCT CUSTOMISATION

The main components of IoT-enabled cloud manufacturing system consist of five different roles based on customers supporting arrangement, as shown in Figure 1.5. Five types of investors in customised and personalised product design and manufacturing are illustrated. Herein, consumers refer to persons or enterprises that require customised and personalised products (Aheleroff et al. 2020). Self-motivated societies and specialised associations fall under third parties who have usual interests regarding customised and personalised products.

The product designer is the foremost contributor to customised and personalised product design, whereas self-motivated societies and specialised associations and consumers will similarly be involved in improved customised and personalised product design with updated customer knowledge and inspired solutions. Quality inspectors and manufacturers are in charge of the dispensation and gathering of products and their corresponding quality inspections (Figure 1.5).

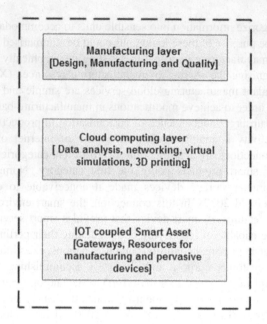

**FIGURE 1.5** Layout of an IOT-Based Cloud Manufacturing System for Product Customisation.

In diverse phases of the customised and personalised product lifespan, the combined contributions from the five categories of societies is essential in product improvement in terms of lower manufacturing cost and lesser production time. The secondary infrastructure is comprised of three main layers, as discussed below (Khan et al. 2020).

### 1.3.2.1 Internet of Things-Based Smart Layer

Material production resources are comprised of digitalised hard resources, such as computer numerical control machines and soft resources such modelling software (e.g. solid works, Auto CAD, etc.). These assets are in the pertinent location to be supervised and are governed by the use of remote controls. For conventional hardware competence, entrenching tiny electronics, viz. sensors and actuators, could integrate them with consequential cyberspace in order to deliver output evidence and collect and implement the instructions. There are also self-directed components that integrally comprise detecting, decision-making and activating units. Altogether, manufacturing resources interact and interchange information respectively with the assistance of wired or wireless communication methods (Zhao et al. 2020). Categorised and circulated entries are meant to enable synchronisation and interoperation amongst various levels of belongings by launching hybrid organisation networks. The entryways could easily help generate upward and downward grading in order to understand multidimensional collaborations. Through integral data investigative procedures and communication comments, smart gateways are proficient in the implementation and transfer jobs collaboratively. The entire manufacturing internet of things progresses under a hybrid governor standard of consolidated ascendancy and circulated independence (Arun 2020).

### 1.3.2.2 Functional Layer

The functioning layer delivers a well-organised execution on several production possessions that could be extensively retrieved in cloud manufacturing facilities through the internet. Initially, enormous quantities of smart resources are inculpated into reserve storage and summarised as elements that allow on-request distribution of facilities. The virtual nature makes it conceivable to enthusiastically distribute possessions from computer-generated units to a sensible unit in order to attain the expanded needs, whereas the purpose of the logic units could be summarised using a service-oriented architecture prototype to serve customers globally (Upadhyay et al. 2020).

The basic facilities delivered by smart entities can be scored to further complex facilities in order to suggest various competencies. Similar to how the cloud working system can use three conveyance simulations to be able to deliver services, possible characteristics of cloud computing services such as analytics storage, visualisation and computing could manufacture services such as robotics, machining, and sensing, among others. With exceptional proficiencies, the manufacturing cloud is a hard foundation for several influential technologies and their sustained services. Due to the improvements in sensory instrument technology and persistent devices, irresistible statistics are composed at exceptional speed. This aids in the examination of diverse data that is developed from various foundations, such as the internet and sensors, in an opportune way and enables faster and improved decision-making in several manufacturing applications (Labati et al. 2020).

Social networks provide a platform in arranging like-minded individuals or expert organisations in a community space for communication and delivery of knowledge, and this produces opportunities beyond the association limits. Social networks could also distribute the most updated data to customers who wish to acquire knowledge from interesting societies by subscription services. In conventional online societies, the customer browses another person's homepage to procure updates, which might be a long process while relevant information is also not receivable. Through modern communication and sharing tools like video sharing, compatible parties can contribute their knowledge with the confidence of complete understanding.

### 1.3.2.3 Manufacturing Application Layer

The manufacturing cloud is a perfect stage for associations once overall data could be appropriately stored, shared and proficiently handled by the cloud's several competencies. In the open Customised and Personalised Product (CPP) design and manufacturing hypothesis, individual necessities can be distributed to social networks where specific designers, companies and professionals could cooperatively be involved in the proposal of theoretical models for customised and personalised products.

The cycle of the real-time data stream between proposal, manufacture and examination permits the final alterations for manufacturing in the product data based on the customers' preferences, and supervising personals can share the modifications to the manufacturing executive for the changing of procedures. However, within the manufacturing division, diverse processing centres could share the conflicts or vital occurrences in a real-time way through IOT in order to make superior judgments adaptively. To protect the excellence of customised and personalised products, an examination is essential and executed when the products have been produced.

Associated quality-related information is collected and taken away from machinery by sensor units that help in assessing the quality of attained products. The data reported in the designated performance of machinery services can be interrelated and evaluated to uphold quality guarantees. When the obtained data confirms the deprivation in machine performance, precautionary actions can be taken prior to facing the machine failures in the view of lessening the possible loss.

## 1.4   ROLE OF CLOUD MANUFACTURING IN PRODUCT LIFECYCLE MANAGEMENT

In present times, the increase in interest for refined and personalised products acts as a major challenge for the industrial sector in meeting the various needs of their customers in order to develop products with better quality but with minimal production time and low costs. Product Lifecycle Management (PLM) is a methodology for facilitating the easy organisation of a product's details. It helps to record and manage the important information about the product from different points of view and aids in plotting the stakeholder details with workflow about the product (Figure 1.6). This fact initiates improvement in better association inside the stakeholders of the inventiveness network. For a successful implementation of the PLM system, it must have the ability to act as an intermediator in confirming the communication among various partners with information on the lifecycle of a product. Cloud manufacturing is an emergent solution established with Website technologies, which are extensively established in modern industries, and herein, users can appeal to services that include manufacturing, product design and management testing that are related to the workflow on the lifecycle of the product (Roda, Macchi, and Albanese 2020; Stark 2020). In some software solutions, manufacturing systems are developed with aids in resources of manufacturing and its function with supportive software and hardware for the customers based on paid services. Based on issues raised in PLM, manufacturing could be considered as an apt resolution to avoiding several problems such as scalability and poor flexibility. At this point, the resources for manufacturing units are released based on the demand and improvisations in the welfare of the enterprise network. Manufacturing-based industries have tackled several problems like the need to deal with customers with a profit of lower margin, customisation of products that have more demand, shorter product lifecycles and geographical dispersion of material suppliers that become complex factors due to various regulations of governments, among others.

In this perspective, manufacturing sectors are struggling to upsurge transparency operations in order to lessen logistics and manufacturing costs, to improvise the

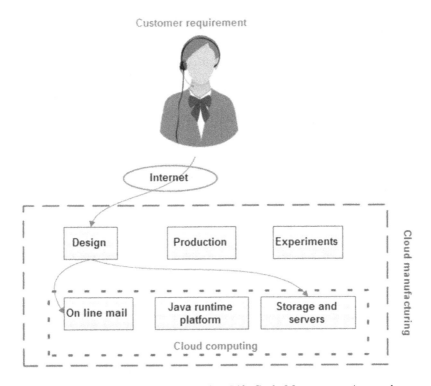

**FIGURE 1.6**    Outline of Cloud-Based Product Life Cycle Management Approach.

quality of the developed product, to reduce the time taken to the marketplace by forming more user-friendly manufacturing lines and also to maximise worth in forthcoming components for shareholders and clients (Liu et al. 2020). Communication and information technologies permit quicker teamwork for leading stakeholders around the comprehensive enterprise in generating, maintaining and using product information. In the CM field, PLM techniques were estimated in addition to assisting the manufacturing sector by collecting the details about the advancement of products, progressions and resource availability through their whole product lifecycle. PLM may consider the business that has advanced from conventional, engineering-based tools into the level of enterprise solution (Cheng et al. 2020).

PLM is in run-through as fruition of components or material data controlling systems (CDCS). CDCS is also considered the main pillar of PLM. It consists of five major operations: first, it is comprised of an information vault that acts as storage for product information that is kept in an organised manner; the second is the information managing modules that help in data access, system administration, integrity and security, recovery and archiving the simultaneous usage of information or data; the third is the workflow managing modules that help in processes outlining and footage; the fourth is the interfacing of the user to support users' requirements; the fifth and final is system interfacing for programs such as computer-aided engineering and draughting.

## 1.5　CLOUD COMPUTING-BASED PLM

A newer class of ICT (i.e. information and communication technologies) stands and services concerned with the virtual platform where prevailing digital tools are disintegrated towards purposeful portions and undergo diverse techniques to progress first-hand industrial applications. These developed applications and models can be integrated into PLM systems to be able to develop scalable, simple and flexible platforms.

These contemporary IT platforms frequently engage the cloud manufacturing standard, which is constructed on the perception of a facility concerned with planning, cloud computing and IoT (i.e. internet of things). CMS is aimed at permitting universal, suitable, request networking admittance in communal groups of convictable possessions that could rapidly be a platform that is unconstrained. Cloud computing might simplify the association among all shareholders across the lifecycle of the product, interrelating companions and permitting transferal for invention documents in low prices and accessible methods. CM suggests facilities through web-assisted computational methods, skills such as programs, platforms and infrastructure provisions. CMS generates cooperative policies attached network strategies and further deals with requests and user-friendly software, IT organisation and platforms justifying the need for trademarked organisation and desktop-based applications. CMS-assisted PLM methods could simplify incorporation in an overabundance of domestic schemes and constituents apt for diverse interior associations and product sellers. Developing a CMS with the assistance of IoT-based criteria will enforce a system-based incorporation (Y. Wang et al. 2020; Ali et al. 2020). The prime benefits of CM are enhanced source consumption, improved dimensional constancy and better association between diverse production resources. CMS has previously revealed its competencies and has the potential to permit PLM to develop exceeding accessibility.

### 1.5.1　THE ROLE OF CM IN RAPID PRODUCT DEVELOPMENT BY ADDITIVE MANUFACTURING

Additive Manufacturing (AM) states that Industry 4.0 allows manufacturing parts that come directly from the modelled, computer-aided draughting documents to be built up product layer by layer, as mentioned in three-dimensional geometrical values. This chapter deals with a detailed view on IOT-cloud-assisted additive manufacturing (Gebhardt 2017; Herzog et al. 2016).

The main motive of a cloud platform is to computerise order processes that are additively manufactured components in a website-based environment. This platform develops a streamlined communicational interface between the user and manufacturing supplier. Furthermore, it assists the supplier in handling the jobs and analysis.

The cloud-based additive manufacturing approach consists of five different phases. First is the workflow phase. In this phase, the client can upload their requirements with accurate geometrical data thanks to the support of online systems such as web browsers (Figure 1.7). Subsequently, the upload details are verified and

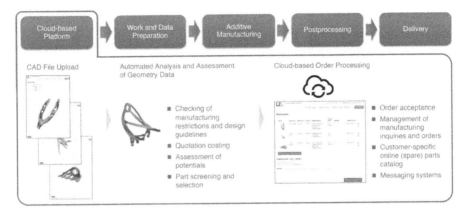

**FIGURE 1.7** Layout for Additive Manufacturing-Based Cloud Platform. Reprinted from Procedia CIRP, Vol. 63, Jan-Peer Rudolph, Claus Emmelmann, A Cloud-Based Platform for Automated Order Processing in Additive Manufacturing, Pages No. 413, Copyright (2017), with permission from Elsevier.

calculated by web servers. Based on the attained results, the client can then order the parts from the manufactures. Moreover, based on design procedures, the manufacturer will subsequently recheck and make the decision to accept or reject the order placed by the client. The second phase deals with geometric data investigations, herein referred to as the geometry of the area, surface and volumes, which are evaluated. The resulting data is employed in the Standard Triangulation Language format (STL) which is the universal file for the additive manufacturing process. Quotation and Part Screening processes are carried out for calculating the cost of the STL file. Next, additive manufacturing can develop many parts in a single job, and the volume consumption for developing jobs is calculated by using the 2D algorithm.

### 1.5.2 CLOUD-BASED PLATFORM FOR ADDITIVE MANUFACTURING

This approach consists of three parts: service providers are placed in the upper part; the CM platforms are placed in the middle; and the client is established in the bottom. Using this, the additive manufacturing resources are summarised as facilities being available to CM platform. Furthermore, service providers are also included to support the commercial processes of cloud additive manufacturing (Figure 1.8). Supplied facility interfaces are also enumerated in the CM platform. Herein, the manufacturing proficiencies supplied by 3D printers are referred to as AM facilities that assist in the implementation of the fabrication process. Furthermore, the CM platform can observe, measure and configure periodically as well as carry out the jobs separately.

There are a few standards and rules for system design that might be used for controlling 3D printers, as well as control factors that will assist in improving the quality of manufacturing by machine learning approach. CM platform

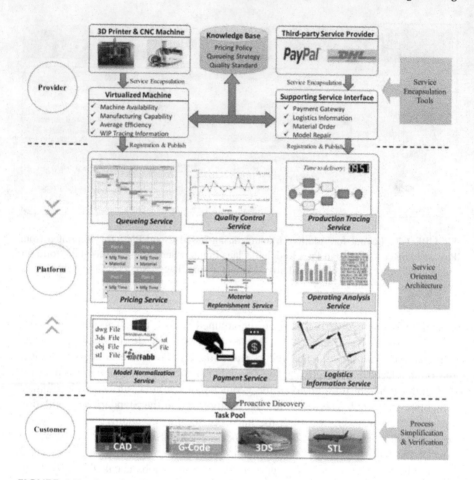

**FIGURE 1.8** Associated Platforms for Cloud-Based Additive Manufacturing Resource Reprinted from Journal of Cleaner Production, Vol. 241, Cheng Qian, Yingfeng Zhang, Yang Liu, Zhe Wang, A Cloud Service Platform Integrating Additive and Subtractive Manufacturing with High Resource Efficiency, Pages No. 4, Copyright (2019), with permission from Elsevier.

consists of two different services that assist in collecting and examining orders and passes the command to additive manufacturing resources. In this case, the client can submit jobs to the job pool. During job submission, the client should gather information on the requirements for their jobs, such as the time of delivery. The entire procedure will then be started when a client submits their manufacturing job. They will upload their geometric design to the CM platform by encoded protocols and also specify the manufacturing necessities for their job. The geometric designs of the components are assigned to manufacturing resources and would be allocated to other resources later. The uploaded geometric designs would be used for manufacturing, and it would be subsequently erased after the job has finished. Conversely, the client can select geometric designs on the CM platform for analysis. The acquired job is then converted to STL file format. This facility would also detect errors and attempt to correct them using cloud computing. The alterations will occur once the STL file format can no longer be evaluated appropriately, and the defects detected in the designs mesh will not alter the actual design. With the assistance of additive manufacturing software such as Repetier-host, such defects can be anticipated. Netfabb, MeshFix and Trinckle are some of the types of software used for defect identification and alteration. The client will be informed if such alterations occur and can be allowed to reconsider only if the altered design files are satisfactory. The job has a manufacturing approach as it fulfils all necessities of geometric design, precision, delivery time and budget etc. In this way, the client can deliver replacements that use diverse manufacturing methods for an identical process. The condition of AM resources that collects by operational examination service and job requirements would be served to line up service, while assisting to initiate the manufacturing methodology and resolving which resources can possibly undertake the job. Afterwards, this pre-processed model will be divided for suitability of the additive manufacturing process. Furthermore, in this step, several manufacturing plans will be presented to the client. These plans consist of various targets and are applied with diverse constraints. The client can, consequently, choose a plan and pay for it affording to the assessment from the pricing service. When the payment has been completed in the pricing platform, the production process guidelines are produced by the queuing service that would be directed to AM resources. The quality control facility is accountable for screening the processes and reconfiguring the essential resources. The production process will be accessible by their client with the assistance of an online service delivered by the manufacturing tracing service. If assembly is needed, then the worker will assemble the component based on the commands from the design, while all components have been produced. Robotic arms workers are also the probable choices when a subtle component is involved. Examinations are monitored in order to conduct the general quality assessed previously supplied the finished components. Real-time logistics information is available to clients by using the platform and assessing the service once they obtain the component (Valilai and Houshmand 2015).

## 1.6 SUMMARY

This chapter has reported extensive updates on recent research based on cloud manufacturing and its significant issues. Based on an undisputable literature survey, we classify the chapter with five major divisions related to its application in various fields that includes production scheduling system, product customisation, product lifecycle management and additive manufacturing. In each division, the interface and application of cloud manufacturing was comprehensively discussed. Overall, cloud manufacturing is considered as the assurance of a newer, contemporary,

18 Manufacturing and Industrial Engineering

collaborative and resource-sharing approach in industrial hypothesis in order to achieve active and wide-reaching sustainable manufacturing.

# REFERENCE

Aheleroff, Shohin, Xun Xu, Yuqian Lu, Mauricio Aristizabal, Juan Pablo Velásquez, Benjamin Joa, and Yesid Valencia. 2020. "IoT-Enabled Smart Appliances under Industry 4.0: A Case Study." *Advanced Engineering Informatics* 43. d 101043.
Ali, Munira Mohd, Mamadou Bilo Doumbouya, Thierry Louge, Rahul Rai, and Mohamed Hedi Karray. 2020. "Ontology-Based Approach to Extract Product's Design Features from Online Customers' Reviews." *Computers in Industry* 116. Elsevier: 103175.
Arun, Achamkulamgara. 2020. "Architecting IOT for Smart Cities." In *Smart Cities in Application*, 141–152. Cham, Switzerland: Springer.
Cheng, Jiangfeng, He Zhang, Fei Tao, and Chia-Feng Juang. 2020. "DT-II: Digital Twin Enhanced Industrial Internet Reference Framework towards Smart Manufacturing." *Robotics and Computer-Integrated Manufacturing* 62. Elsevier: 101881.
Dong, Tingting, Fei Xue, Chuangbai Xiao, and Juntao Li. 2020. "Task Scheduling Based on Deep Reinforcement Learning in a Cloud Manufacturing Environment." *Concurrency and Computation: Practice and Experience*. 32 (5). Wiley Online Library: e5654.
Gebhardt, Andreas. 2017. *Additive Fertigungsverfahren: Additive Manufacturing Und 3D-Drucken Für Prototyping-Tooling-Produktion*. Munchen, Germany: Carl Hanser Verlag GmbH Co KG.
Herzog, Dirk, Vanessa Seyda, Eric Wycisk, and Claus Emmelmann. 2016. "Additive Manufacturing of Metals." *Acta Materialia* 117. Elsevier: 371–392.
Hu, Rongbo, Thomas Linner, Julian Trummer, Jörg Güttler, Amir Kabouteh, Katharina Langosch, and Thomas Bock. 2020. "Developing a Smart Home Solution Based on Personalized Intelligent Interior Units to Promote Activity and Customized Healthcare for Aging Society." *Journal of Population Ageing* 13. Springer: 1–24.
Hu, Yanjuan, Lizhe Wu, Chao Shi, Yilin Wang, and Feifan Zhu. 2020. "Research on Optimal Decision-Making of Cloud Manufacturing Service Provider Based on Grey Correlation Analysis and TOPSIS." *International Journal of Production Research* 58 (3). Taylor & Francis: 748–757.
Jin, Xinjuan, Yu Shiqiang, Zheng Pai, Liu Quan, and Xu Xun. 2018. "Cloud-based Approach for Smart Product Personalization." *Elsevier:Procedia CIRP* 72. 922–927.
Khan, W. Z., M. H. Rehman, H. M. Zangoti, M. K. Afzal, N. Armi, and K. Salah. 2020. "Industrial Internet of Things: Recent Advances, Enabling Technologies and Open Challenges." *Computers & Electrical Engineering* 81. Elsevier: 106522.
Labati, Ruggero Donida, Angelo Genovese, Vincenzo Piuri, Fabio Scotti, and Sarvesh Vishwakarma. 2020. "Computational Intelligence in Cloud Computing." In *Recent Advances in Intelligent Engineering*, 111–127. Cham, Switzerland: Springer.
Laili, Yuanjun, Sisi Lin, and Diyin Tang. 2020. "Multi-Phase Integrated Scheduling of Hybrid Tasks in Cloud Manufacturing Environment." *Robotics and Computer-Integrated Manufacturing* 61. Elsevier: 101850.
Li, Tianyang, Ting He, Zhongjie Wang, and Yufeng Zhang. 2020. "SDF-GA: A Service Domain Feature-Oriented Approach for Manufacturing Cloud Service Composition." *Journal of Intelligent Manufacturing* 31 (3). Springer: 681–702.
Li, Xiaobin, Zhiwei Fang, and Chao Yin. 2020. "A Machine Tool Matching Method in Cloud Manufacturing Using Markov Decision Process and Cross-Entropy." *Robotics and Computer-Integrated Manufacturing* 65. Elsevier: 101968.

Li, Bo-Hu, Lin Zhang, and Xudong Chai. 2010. "Introduction to Cloud Manufacturing." *ZTE Communications* 16 (4): 5–8.

Lin, Yang-Kuei, and Chin Soon Chong. 2017. "Fast GA-Based Project Scheduling for Computing Resources Allocation in a Cloud Manufacturing System." *Journal of Intelligent Manufacturing* 28 (5). Springer: 1189–1201.

Liu, Yongkui, Xun Xu, Ananth Srinivasan, and Lin Zhang. 2017. "Enterprises in Cloud Manufacturing: A Preliminary Exploration." In *ASME 2017 12th International Manufacturing Science and Engineering Conference Collocated with the JSME/ASME 2017 6th International Conference on Materials and Processing.* American Society of Mechanical Engineers Digital Collection.

Liu, Yang, Yingfeng Zhang, Shan Ren, Miying Yang, Yutao Wang, and Donald Huisingh. 2020. "How Can Smart Technologies Contribute to Sustainable Product Lifecycle Management?" *Journal of Cleaner Production* 249. Elsevier: 119423.

Ma, Jing, Hua Zhou, Changchun Liu, E. Mingcheng, Zengqiang Jiang, and Qiang Wang. 2020. "Study on Edge-Cloud Collaborative Production Scheduling Based on Enterprises With Multi-Factory." *IEEE Access* 8. IEEE: 30069–30080.

Mont, Oksana K. 2002. "Clarifying the Concept of Product–Service System." *Journal of Cleaner Production* 10 (3). Elsevier: 237–245.

Mourad, Mohamed H., Aydin Nassehi, Dirk Schaefer, and Stephen T. Newman. 2020. "Assessment of Interoperability in Cloud Manufacturing." *Robotics and Computer-Integrated Manufacturing* 61. Elsevier: 101832.

Mourtzis, Dimitris, Ekaterini Vlachou, Nikolaos Milas, and George Dimitrakopoulos. 2016. "Energy Consumption Estimation for Machining Processes Based on Real-Time Shop Floor Monitoring via Wireless Sensor Networks." *Procedia CIRP* 57. Elsevier: 637–642.

Pei Breivold, Hongyu. 2020. "Towards Factories of the Future: Migration of Industrial Legacy Automation Systems in the Cloud Computing and Internet-of-Things Context." *Enterprise Information Systems* 14 (4). Taylor & Francis: 542–562.

Porter, Michael E., and James E. Heppelmann. 2014. "How Smart, Connected Products Are Transforming Competition." *Harvard Business Review* 92 (11): 64–88.

Roda, Irene, Marco Macchi, and Saverio Albanese. 2020. "Building a Total Cost of Ownership Model to Support Manufacturing Asset Lifecycle Management." *Production Planning & Control* 31 (1). Taylor & Francis: 19–37.

Simeone, Alessandro, Bin Deng, and Alessandra Caggiano. 2020. "Resource Efficiency Enhancement in Sheet Metal Cutting Industrial Networks through Cloud Manufacturing." *The International Journal of Advanced Manufacturing Technology* 107. Springer: 1–21.

Stark, John. 2020. "PLM, Techniques and Methods." In *Product Lifecycle Management (Volume 1)*, 309–333. Cham, Switzerland: Springer.

Tao, Fei, Lin Zhang, V. C. Venkatesh, Y. Luo, and Ying Cheng. 2011. "Cloud Manufacturing: A Computing and Service-Oriented Manufacturing Model." *Proceedings of the Institution of Mechanical Engineers, Part B: Journal of Engineering Manufacture* 225 (10). Sage Publications: 1969–1976.

Upadhyay, Shivansh, Shashwat Kumar, Sagnik Dutta, Ajay Kumar Srivastava, Amit Kumar Mondal, and Vivek Kaundal. 2020. "A Comprehensive Review on the Issues Related to the Data Security of Internet of Things (IoT) Devices." In *Intelligent Communication, Control and Devices*, 727–734. Singapore: Springer.

Valilai, Omid Fatahi, and Mahmoud Houshmand. 2015. "Depicting Additive Manufacturing from a Global Perspective; Using Cloud Manufacturing Paradigm for Integration and Collaboration." *Proceedings of the Institution of Mechanical*

*Engineers, Part B: Journal of Engineering Manufacture* 229 (12). SAGE Publications: 2216–2237.

Wan, Changcheng, Hualin Zheng, Liang Guo, Xun Xu, Ray Y. Zhong, and Fu Yan. 2020. "Cloud Manufacturing in China: A Review." *International Journal of Computer Integrated Manufacturing* 33. Taylor & Francis: 1–23.

Wang, Ke. 2020. "Migration Strategy of Cloud Collaborative Computing for Delay-Sensitive Industrial IoT Applications in the Context of Intelligent Manufacturing." *Computer Communications* 150. Elsevier: 413–420.

Wang, Yankai, Shilong Wang, Bo Yang, Lingzi Zhu, and Feng Liu. 2020. "Big Data Driven Hierarchical Digital Twin Predictive Remanufacturing Paradigm: Architecture, Control Mechanism, Application Scenario and Benefits." *Journal of Cleaner Production* 248. Elsevier: 119299.

Wang, Z., J. H. Zhang, and Y. Q. Qi. 2017. "Job Shop Scheduling Method with Idle Time in Cloud Manufacturing." *Control and Decision* 32 (5): 811–816.

Wu, Lei, and Chengwei Yang. 2010. "A Solution of Manufacturing Resources Sharing in Cloud Computing Environment." In *Cooperative Design, Visualization, and Engineering*, Yuhua Luo Ed. Heidelberg, Germany: Springer247–252.

Xia, Ting, Wei Zhang, W. S. Chiu, and Changqiang Jing. 2020. "Using Cloud Computing Integrated Architecture to Improve Delivery Committed Rate in Smart Manufacturing." *Enterprise Information Systems*. Taylor & Francis: 1–20.

Xu, Xun, Lihui Wang, and Stephen T. Newman. 2011. "Computer-Aided Process Planning–A Critical Review of Recent Developments and Future Trends." *International Journal of Computer Integrated Manufacturing* 24 (1). Taylor & Francis: 1–31.

Yang, Chen, Shulin Lan, Weiming Shen, George Q. Huang, Xianbin Wang, and Tingyu Lin. 2017. "Towards Product Customization and Personalization in IoT-Enabled Cloud Manufacturing." *Cluster Computing* 20 (2). Springer: 1717–1730.

Yang, Yefeng, Bo Yang, Shilong Wang, Tianguo Jin, and Shi Li. 2020. "An Enhanced Multi-Objective Grey Wolf Optimizer for Service Composition in Cloud Manufacturing." *Applied Soft Computing* 87. Elsevier: 106003.

Zhang, Lin, H. Guo, Fei Tao, Y. L. Luo, and N. Si. 2010. "Flexible Management of Resource Service Composition in Cloud Manufacturing." In *2010 IEEE International Conference on Industrial Engineering and Engineering Management*, 2278–2282. IEEE.

Zhang, Lin, Yong-Liang Luo, Wen-Hui Fan, Fei Tao, and Lei Ren. 2011. "Analyses of Cloud Manufacturing and Related Advanced Manufacturing Models." *Computer Integrated Manufacturing Systems* 17 (3): 458–468.

Zhang, Lin, Yongliang Luo, Fei Tao, Bo Hu Li, Lei Ren, Xuesong Zhang, Hua Guo, Ying Cheng, Anrui Hu, and Yongkui Liu. 2014. "Cloud Manufacturing: A New Manufacturing Paradigm." *Enterprise Information Systems* 8 (2). Taylor & Francis: 167–187.

Zhao, Zhiheng, Mengdi Zhang, Gangyan Xu, Dengyin Zhang, and George Q Huang. 2020. "Logistics Sustainability Practices: An IoT-Enabled Smart Indoor Parking System for Industrial Hazardous Chemical Vehicles." *International Journal of Production Research* 58. Taylor & Francis: 1–17.

Zhou, Longfei, Lin Zhang, and Yajun Fang. 2020. "Logistics Service Scheduling with Manufacturing Provider Selection in Cloud Manufacturing." *Robotics and Computer-Integrated Manufacturing* 65. Elsevier: 101914.

Zhu, Xiaobao, Jing Shi, Fengjie Xie, and Rouqi Song. 2020. "Pricing Strategy and System Performance in a Cloud-Based Manufacturing System Built on Blockchain Technology." *Journal of Intelligent Manufacturing* 31. Springer: 1–18.

# 2 Bonding and Cladding of Composite Materials

*K. Ravi Kumar and T. Pridhar*

## CONTENTS

## 2.1 INTRODUCTION

A composite material is made by two or more different materials where one acts as a matrix and the other as the reinforcement. Composites have superior physical and mechanical properties than its base materials (Ravi Kumar et al. 2018). Conventionally, the composites are classified based on the three types of matrices, namely metal matrix composites (MMC), polymeric matrix composites (PMC) and ceramic matrix composites (CMC) (Ravi Kumar et al. 2018; Altenbach et al. 2018; Venkatesh et al. 2019). Natural to high-tech fibres, and even nanofibres, are used as reinforcements. Over the last two decades, there was a rapid development in the area of advanced materials and technologies that include biomaterials, nanomaterials and smart composites which are fabricated with cutting-edge technologies. Nowadays, researchers nowadays working on advanced composites materials as an alternative to traditional components. In the recent past, research, innovation and applications of composite materials have increased due to their superior multiperformance characteristics. Cladding is one of the important techniques used to fabricate a composite by layering dissimilar materials (Ha and Hong 2013).

This process does not use any fillers or adhesive materials. Rather, bonding takes place with the help of extreme pressure in the presence or absence of heat. Cladding is capable of producing single clad (two layers), double-clad (three layers) and even up to seven layers or more depending on the materials.

Cladding was developed in 1930. Early applications include materials-handling equipment, heat exchangers, tanks, processing vessels and storage equipment, among others. Cladding a certain kind of metallic bonding generally involves a three-stage process, namely: (1) establishing physical contact (2) surface activation and (3) metal interaction during hot or cold process (Li et al. 2008). The properties of clad materials are influenced by the stacking structure of different materials, selection of material composition, thickness of material surfaces, interface structure and processing methods. While selecting different materials for cladding in addition to the properties of individual materials, the interfacial reaction and interface structure between the two materials are of utmost importance. The properties of the clad composite plates depend on the thickness of the constituent metal layers. The subsequent sections shed light on various bonding and cladding techniques for composites.

## 2.2 ROLL BONDING

### 2.2.1 INTRODUCTION

Multilayer composites have attracted increasing attention in industries because of their superior magnetic, mechanical and electrical properties. Among available technologies, such as explosive welding, electroplating and transfer welding, roll bonding has attracted researchers due to their specific advancements compared to the other methods. Cold roll bonding is a widely used solid-state bonding manufacturing process that has the capability of joining both similar and dissimilar metals (Jamaati and Toroghinejad 2011). Roll bonding is a solid-phase process in which the metals are bonded by rolling at an applied pressure. In roll bonding, two or more metals in the form of sheets, strips or plates are stacked over each other. Next, they are then roll bonded in order to achieve a suitable bonding among the metal strips. The performance of roll bonded composites is influenced by parameters like strength and the hardness of the workpiece material, deformation, applied pressure, time and temperature of roll bonding, post heat treatment, rolling speed, thickness and surface preparation.

### 2.2.2 WORKING PRINCIPLE

In cold roll bonding, the bonding of metallic sheets or plates occurs by plastic deformation in solid states due to rolling (Jamaati and Toroghinejad 2011). The schematic representation of roll bonding is illustrated in Figure 2.1. Roll bonding consists of two or more metals in the form of plates, sheets or strips placed over each other and stacked together by spot welding, riveting or clamping. Next, it is then passed on through a pair of rolls and the materials undergo deformation and establish a solid-state bond between each other. The high pressure of the roller during bonding causes a high thickness reduction of the materials. During deformation, an expansion of the contact surfaces takes place, and this eventually breaks up the surface layer and the thin film particles. This causes the filling of the materials into the cracks and results in bonding. Due to the high reduction in thickness, the surface of the materials generates

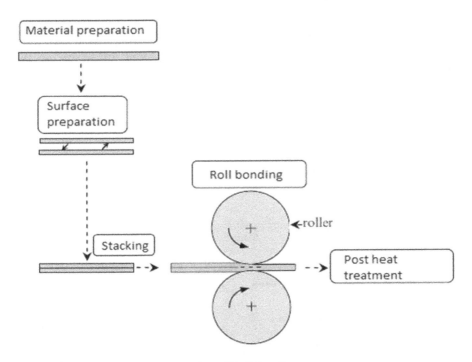

**FIGURE 2.1** Schematic Representation of Roll Bonding.

heat to a larger extent and bonding takes place among the materials. Bonding occurs by the atomic affinity and the interfacial mechanical bonding between the materials. A maximum of 50%, or sometimes even higher, reduction in thickness can be achieved in a single rolling pass. Prior to bonding, it is essential to clean the surfaces of the workpiece material and remove the surface layers. It is achieved by degassing and scratching the surfaces in order to remove the contaminants and surface oxides that might interfere and affect bonding. Preheating of the workpiece has to be carried in the case of hot working, and in terms of cold working, the material is used in atmospheric heat condition. Generally, annealing is carried out after rolling to be able to improve the bonding strength at the interface between the materials and also to reduce the strain and hardness of the materials and the intermetallic layers formed during bonding. The layered composite can be converted into the required shape by using conventional manufacturing processes like forming.

### 2.2.3 ACCUMULATIVE ROLLING BONDING

Recently, studies were carried out in producing ultra-fine grains (UFG) with a diameter of fewer than 1 μm and with the help of an intense plastic straining process by severe plastic deformation (SPD). Accumulative Rolling Bonding is a cutting-edge technique that has the capability to produce components with fine and ultra-fine-grains by SPD (Ghalehbandi et al. 2019). The basic fundamental experimental steps of accumulative roll bonding are the same as the roll bonding process. The product obtained during the first rolling cycle is cut into two similar sheets and is subsequently placed one over the other, stacked and rolled again as shown in Figure 2.2. After completing the first pass, the rolled specimens are cut into two equal halves, cleaned and degreased and wire-brushed before stacking. The sacked component then undergoes preheating before the next rolling process. Normally, the two sheets are rolled to a 50% reduction during each pass. This process can be repeated up to about ten times or even more depending on the other parameters. The accumulative roll

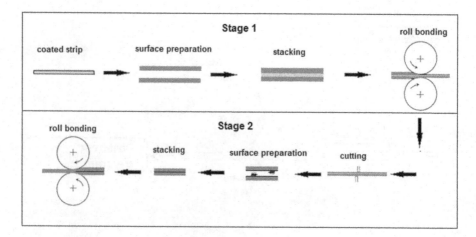

**FIGURE 2.2**   Stages of Accumulative Roll Bonding.

bonding processes can achieve a high-strained finished product with ultra-fine grains with or without preheating. Through the accumulative roll, bonded components normally have a strength that is 3–5 times higher than the parent metals.

## 2.2.4 Mechanism of Roll Bonding

Researchers propose that the mechanism of cold roll bonding is governed by four theories, namely: the (i) thin-film theory, (ii) energy barrier theory, (iii) diffusion bonding theory and (iv) joint recrystallisation theory (Da Silva et al. 2014). The film theory is the primary aspect that influences cold bonding. According to this theory, when two clean metal surfaces are brought into contact under a certain amount of pressure, the material expands and a further increase in deformation beyond a certain limit breaks the surface layer and leads to the formation of cracks. Additional rolling causes extrusion of the base material through the cracks and leads to cold bonding. The film theory is based on two bonding mechanisms: (i) the brittle cover layers of the surfaces formed by scratching become fractured due to the surface expansion during rolling leading to the surface expansion, and the material extrudes into the crack surfaces and creates a metallic bond; (ii) in the absence of cover layer, local thinning of the film comprising of oxides and other contaminants takes place due to surface expansion and reaches threshold value required for bonding that may vary based on the condition of the metal surface (Da Silva et al. 2014). The relative movement between the two surface layers is exposed to high friction during rolling. Next, the scratch-brushed layers then get fractured and form debris of sprinkled blocks. As the base metal undergoes rolling, mechanical interlocking simultaneously takes place. The brushed, work-hardened surface splits and forms coherent blocks along the interfacial surface. During rolling, the parent material becomes exposed between the harder blocks and the material extrudes into the hardened layers (Sardar et al. 2015). The bonding ability of the material is because of differences in plastic properties, stacking fault energy and hardness ratio. The energy barrier theory is concerned with factors like recrystallisation that act as a barrier for bonding clean surfaces. The reactions between metal bonds depend on the three major stages: (i) establishment of physical contact, (ii) energising of the contact surfaces and (iii) interaction between the metallic bonding (Li et al. 2008).

## 2.2.5 Process Parameters

The important process parameters of roll bonding are: the type and nature of base metal, the bonding temperature, rolling pressure, bonding time, surface preparation, layer thickness, number of layers, rolling speed, post- and pre-heat treatment and stacking sequence, among others. The reduction in thickness decreases the threshold deformation and improves bonding strength. Pre-annealing and post-annealing treatments increase the bonding strength of the metals, and the impact is higher in post-rolling annealing treatment compared to that of pre-rolling annealing treatment (Jamaati and Toroghinejad 2010). An increase in initial thickness normally reduces the durability of the interlayer bond. The contact time and temperature developed at the rolling zone significantly influence the rolling speed.

Higher speed generally increases the temperature and decreases the contact time of the metal interfaces; hence, it is required to select the optimum rolling speed. Transverse rolling decreases the bonding strength when compared to that of the actual rolling direction (Jamaati and Toroghinejad 2011). Microhardness near the welded interface is higher due to the elongation of grains perpendicular to the welding direction. Deformation hardening ensures the higher hardness of the cladded plates of composites than that of their parent material; the maximum hardness value remains normally high at the interface area of cladded metals.

Tensile strength and impact strength of the clad composites are higher than that of the base material due to high impact toughness (Dhib et al. 2016). The threshold reduction for bonding for cold cladding is greater compared to that of the hot rolling process. The presence of strain-induced hard layers plays a significant role in improving the tensile tenacity. The post-annealing treatment is used to overcome problems in mechanical properties due to softening and formation of the intermediate layer at high temperatures that reduce the tensile strength (Akramifard et al. 2014). An increase in the number of cycles in accumulative roll bonding influences the bonding due to necking, distribution of reinforcing phase and the strain originated due to friction. Strength and microhardness of the clad composites increase with the upsurge in the number of cycles, while the elongation decreases compared to that of the base material. In addition to the intermetallic layer, certain non-thermodynamically intermixing is also observed in accumulative roll bonding (Mozaffari et al. 2010).

## 2.3 FRICTION SURFACE CLADDING

### 2.3.1 INTRODUCTION

Industries like aerospace, automotive and shipbuilding make use of technologies and processes that enhance surface properties such as wear resistance, corrosion resistance, chemical resistance and hardness of the manufactured components. The surface is coated with layers that are different compared to the base metal. Surface modification techniques like physical vapour deposition, chemical vapour deposition, shielded metal arc cladding, thermal spraying, laser beam cladding and friction surfacing are used in industries. Friction surfacing is an additive manufacturing technique that utilises a rotating consumable rod to generate frictional heat between the substrate material and the consumable rod to coat thin metallic layers on a substrate. Friction surface cladding is used mainly for fabricating wear and corrosion-resistant coatings and for surface recovery of worn engineering components. The functional performance of the component is improved by surface modification processes. The temperature generated during friction surfacing is significantly much less than the melting point of the substrate; hence, fusion defects and defects of the heat-affected zone are reduced (Liu et al. 2016).

### 2.3.2 WORKING PRINCIPLE

Friction surfacing cladding is a solid-state surface engineering process where a consumable rod is used to clad the material by its plastic deformation. It is an energy-efficient process as well as an environmentally friendly surface cladding technique. The consumable rod made of the clad material is pressed against the workpiece substrate under an axial load (Figure 2.3) that causes the generation of heat due to friction at the area of contact between the consumable rod and substrate plate. This frictional heat develops a viscoplastic boundary layer at the tip of the consumable rod. As the rod moves along the surface of the substrate, the viscoplastic material starts flowing to the neighbouring area of the interface. Moreover, the heat generated along with pressure triggers the interdiffusion process and forms a sturdy bond at the interface between the deposits and the substrate. The performance of friction surface cladding depends on major parameters like translation speed of substrate, tool gap distance, tool tilt angle, the diameter of the consumable rod, applied pressure, the surface roughness of the substrate and rotation speed. The resulting friction depends on the heat generated by interfacial friction, plastic deformation and works on temperature below the fusion of the substrate and leads to the deposition of fine-grained particles on the substrate. Friction surfacing is also referred to as a coating process done to improve the strength, wear and corrosion resistance.

Friction surfacing produces clad layers with a defined thickness and width on a substrate. The cladding of materials takes place due to two reasons: (1) the clad material deposits at the top of the substrate and forms a non-intermixed layer that contains solely the clad material; (2) portion of the substrate becomes mixed with the rod to form a clad layer. The heat in friction surface cladding is generated by (1) the friction between substrate and rotating consumable rods and (2) the friction at the interface of the tool with the clad layer on the substrate. The interface between the tool and the substrate determines the clad layer thickness. During the process, the clad material gets softened and deforms at elevated temperatures, while the substrate

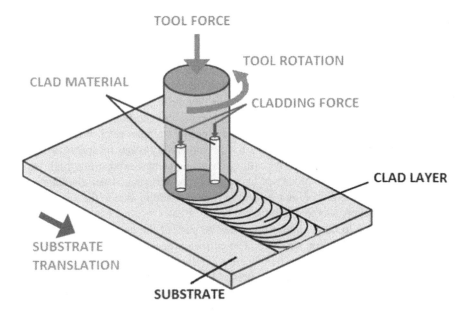

**FIGURE 2.3** Schematic Representation of Friction Surface Cladding Process.

remains substantial. Porosity, severe plastic deformation and subsequent loss of clad material may be the major issues encountered in friction cladding.

### 2.3.3 PROCESS PARAMETERS

The quality of uniform width and thickness of friction clad composites depend on the shape of the ring formed at the tip of the rod as well as the diameter of the rod. The deposition of the flat shape of the rod ring influences a uniform width and thickness (Liu et al. 2016). At low temperatures, surface defects and improper mixing of the clad layer with the substrate occur. The formation of a heat-affected zone increases with the rise in tool temperatures beyond a certain limit and simultaneously lowers the hardness of the substrates (Liu et al. 2016).

Axial force, spindle speed and table traverse speed are the major factors that influence the cladding of Al alloy and mild-steel to be done by friction surfacing. The coating thickness decreases while expanding in coating width. The thickness and width of the coating depend on the torque developed during cladding. The width and thickness are also influenced by the effect of interaction among the traverse speed, axial force and the spindle speed. The traverse speed heightens the influence of coating width and thickness. A higher interaction between the axial force and table speed also improves the coating width and thickness (Kumar et al. 2015).

The process temperature highly affects the quality of the deposited layers. At low process temperature, discontinuous clad layers are formed on the substrate, whereas high process temperature leads to the mixing of the substrate with the clad layer, thus resulting in the formation of a continuous clad layer. The quality of the clad layer primarily depends on the plastic deformation and frictional heat acquired during the process that, in turn, depends on the tool rotation rate and the amount of supply of the clad material. The process temperatures and heat-affected zone heavily influence the bending strength of clad composites (Badheka and Badheka 2017). The cooling rate of the material after and during processing influences the grain size deposited on the substrate. The microstructural characteristics of the clad materials depend on the rotational speed and traverse speed. The bond strength becomes weak as the traverse speed increases beyond a limit (Saw et al. 2018).

A wide range of materials like aluminium, magnesium and titanium alloys can be deposited by friction surfacing on alloys and types of stainless steel. Friction surfacing is suitable for joining materials that have compatibility issues incapable of being processed by fusion processes. The casting mechanisms and exposure to high temperature observed in other processes like plasma arc surfacing, laser cladding and shielded metal arc welding are not evident in friction cladding due to the viscoplastic deformation. Friction surfaced coatings produce greater hardness and fine microstructure compared to other similar processes because of the chemical homogeneity and absence of solidification structure. The corrosion resistance of friction surfaced coatings is also high due to the temperature involved and the absence of substrate melt that consequently eliminates the presence of oxide layer. The major problem associated with friction surfacing is the poor bonding located at the coating edge. The revolving flash developed at the tip of the consumable rod decreases the mass transfer efficiency and reduces the bonding performance (Gandra et al. 2014).

## 2.4 LASER CLADDING

### 2.4.1 INTRODUCTION

Laser cladding is a metal deposition process and is used commercially to repair high-value materials (Weng et al. 2015). Laser cladded composites possess dense microstructure, strong metallurgical bonding with the substrates and other notable properties. In laser cladding, a nozzle is used to blow powder on the work surface, which becomes deposited and melted on the material substrate, and a track is formed. Laser cladding is suitable for combinations such as St/St, Cu/St, Al/St and Al/Cu, titanium alloys, etc. Studies were also done by using a combination of ceramics and metals. Repeated heating and cooling cycles during layers building ensure solidification of the layers and heat treatment of the solidified material take place after melting.

### 2.4.2 WORKING PRINCIPLE

Laser cladding employs a high-powered laser beam as the primary heat source for the deposition of a thin layer of coating material to the parent metal (Weng et al. 2015; Paydas et al. 2015). The coating material can be pre-placed on the substrate by powder injection or by wire feeding. Among the above, the powder injection method is widely used due to its capabilities such as higher energy efficiency, better reproducibility, ability to clad with any parent metal with controlled thickness. In this process, the powder particles are deposited on the moving substrate to form a layer of thickness ranging from 50 µm to 2 mm. The working principle is illustrated in Figure 2.4. Initially, the cladding powder is preheated to 180°C in the machine head by utilising the laser beam, before being absorbed by the melt pool. Once the melt pool solidifies, a strong metallurgical bond is formed with a highly dense structure between the welding bead and moving substrate, as shown in Figure 2.5. The laser beam which bears superior localised energy dispersion melts a rather thin layer of the workpiece with a minimum dilution of the substrate within the bead. Solidification in the range of $10^{-2}$ to $1$ ms$^{-1}$ can be achieved by laser cladding. A sufficient number of adequate weld beads are placed for uniform dispersion of cladding powder and complete coverage over the workpiece.

There are two different methods for the production of components by laser cladding, namely: 1. two-step laser cladding and 2. one-step laser cladding.

#### 2.4.2.1 Two-Step Laser Cladding

In this method, clad powders are preplaced on the substrate. Chemical binders are usually mixed with the powder to ensure cohesion between the substrate and the cladding powder. The passing of laser beam on the top surface of the substrate creates a melt pool (Figure 2.5). Finally, a fusion bonding is created. The prevention of powder particles from the gas that flows during melting is essential. This can be achieved with the help of a binder mixed with the powder that ensures proper cohesion with the substrate. Moreover, there is a possibility for the formation of porosity in the clad formed during the evaporation process, and this has to be monitored. Proper care must be taken to eliminate the following during

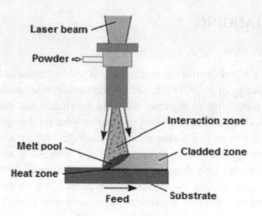

**FIGURE 2.4** Schematic Diagram of Laser Cladding Process.

the process: (i) fusion bond formation due to infiltration of heat into the substrate, (ii) expansion of melt pool due to the conduction of heat by the substrate at the interface and (iii) formation of melt cavity on the surface of the powder due to laser radiation.

### 2.4.2.2 One-Step Laser Cladding

In this process, clad powder particulates are pre-heated in the cladding head. It is then passed into the thermal zone to form a cladding layer (Figure 2.6).

The process parameters involved in laser cladding are listed in Table 2.1.

Laser cladding is suitable for a variety of combinations, such as St/St, Cu/St, Al/St and Al/Cu, titanium alloys, etc. Studies were also done using a combination of ceramics and metals. In addition to the excellent process control of laser cladding, the concentrated heat input due to the laser is also the major factor that influences high accuracy. The large temperature developed during the process leads to the formation of high thermal stresses that generates opportunities for the formation of delamination, cracking and large bending distortion. The thermal contraction that takes place during and after the coating process also paves ways

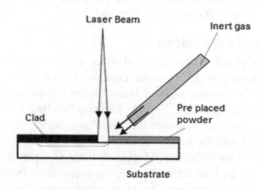

**FIGURE 2.5** Two-Step Laser Cladding.

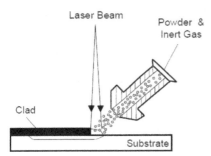

**FIGURE 2.6**   One-Step Laser Cladding.

for the formation of tensile stresses which can be managed by proper cooling process. The process parameters of laser cladding like energy spatial distribution, temperature distribution in the powder particles along the substrate and energy absorbed, among other things, are influenced by the relation between the laser radiation with the powder jet.

Ti-6Al-4V titanium alloy coated with metal matrix composite coatings by laser cladding using SiC, $B_4C$ and $Y_2O_3$ as clad materials revealed that the coatings are mainly reinforced by materials like NiTi, $TiB_2$, $Cr_7C3$, CoTi, $CoTi_2$, $Cr7C_3$, TiC, TiB and $Ti_5Si_3$ (Paydas et al. 2015). There occurs a chemical reaction during the laser cladding process. The hardness of the cladding coatings has increased

## TABLE 2.1
## Laser Cladding Process Parameters

| INPUT | |
|---|---|
| **Laser** | **Material and Its Feeding** |
| • average power | • substrate geometry |
| • spot Size | • clad powder size |
| • wavelength | • composition |
| • motion velocity | • clad powder feed rate |
| | • nozzle Specifications |

| PROCESS |
|---|
| • absorption |
| • melt pool dynamics |
| • melt pool interaction |
| • solidification |

| OUTPUT |
|---|
| • geometry |
| • microstructure |
| • hardness |
| • dilution |
| • surface roughness |
| • cracks |

3–4 times compared to the Ti-6Al-4V substrate; also, the wear resistance of the clad surface increased up to ten times when compared to that of the substrate. Ti-6Al-4V deposited on a machined substrate indicates the formation of some gas porosities along the surface. The clad material deposits exhibited greater strength and ductility than that of the substrate. This can be managed by acquiring a good thermo-metallurgical scheme at the interface. The microstructure and hardness are also influenced by the rate of cooling (Paydas et al. 2015).

Ni- and Fe-based alloy powders, added into FeCrBSi alloy powder that have high Cr content exhibited two different phases, namely, the hard and soft phases. The hard phases were of carbide and the soft ductile phases were of austenite Fe and Ni. Smooth cladding layers with excellent toughness free from cracks, macroscopic pores and void were achieved by laser cladding (Wang et al. 2014).

The microstructures of the clad layers formed by Ni-Cr layers exhibited cellular dendritic structure, where the nickel-chrome dendrites were surrounded by hard precipitates. Formation of boron carbide surface by Ni-boron particle heightened the wear resistance of the surface. The addition of chromium, as well as Molybdenum and tungsten, to the powder magnified corrosion resistance (Ray et al. 2014).

Laser cladding is primarily used in areas of producing functionally graded coating, repair, remanufacturing and near-net-shape production. Trends in laser cladding also focus on multifunctional coatings having properties like the combination of thermal resistance-low friction coefficient, wear-corrosion resistance, high electrical conductivity-abrasion resistance, etc. (Doubenskaia et al. 2015). The worn surfaces of coating (Ni3Ti, Ni-Cr-Fe, $Cr_2Ti$, $TiB_2$ and TiC) and substrate exhibit abrasive wear. The ploughing grooves formed on the substrate were deep and dense, whereas the track on the coating was shallow (Han et al. 2018).

Inconel 625/$Cr_3C_2$ composite coatings produced by laser cladding using $Cr_3C_2$ powders resulted in crack-free composite coatings with a unique structural feature. Impact angles at 30° and 90° of the powders produced composites with higher erosion resistance due to the proper mechanical interlocking with the matrix and the reinforcement that resists trans granular cracking of the carbide phase that occurs during normal impingement (Janicki, 2017).

Laser cladding also has advantages over conventional coating processes, such as plasma spraying and arc welding. It has the ability to produce excellent coating with minimal distortion and the least possible dilution, is free of microcracks and pores, is on a stress concentration-free interface and has better surface quality (Shepeleva et al. 2000).

Applications of laser cladding include:

- coating and creating a wear and corrosion resistant surface for the components like turbine blades, engine valve seats, hydraulic pump components, bearing surfaces, seal surfaces, etc.
- repair and refurbishment of components used to retrieve areas damaged due to prolonged usage
- rapid fabrications of complex parts, cutting tools and dies.

## 2.5  MICROWAVE CLADDING

### 2.5.1  INTRODUCTION

Currently, many types of unconventional processing techniques are used for the manufacturing and fabrication of a wide variety of engineering materials. Quality, cost and sustainability are always the major factors in determining the appropriateness of the process. Manufacturing using microwave radiations results in superior physical, mechanical and metallurgical properties along with attaining cost-efficiency and product quality. In traditional processing techniques, the substrate surface is heated and transmitted to the materials by conduction, convention and radiation. In microwave cladding, heating is done up to the atomic level, which leads to volumetric heating (Gupta and Sharma 2011; Sharma and Gupta 2012; Gupta and Sharma 2014). In this procedure, the electromagnetic energy becomes transformed into heat in the core centre of the substrate, which then proceeds towards the outer surface. Microwave cladding is developed by irradiating the substrate that is covered with powder by using microwaves at frequencies of approximately 2.45 GHz and power of around 900 W. This is employed in materials like Ni-SiC based composite claddings on stainless steel.

### 2.5.2  WORKING PRINCIPLE

In the microwave cladding process, the substrate is initially washed thoroughly in an ultrasonic bath. The clad particulates, with an average size of 40 microns, are preheated at 100°C for 24 hours in a muffle furnace to remove the moisture content. The clad particles are placed on the substrate in uniform dispersion. Figure 2.7 presents the working principle of microwave cladding. Microwave interaction is purely a substrate-dependent function. At room temperature, the clad particles cannot directly interact with microwave radiation, but rather, tries to reflect it. To

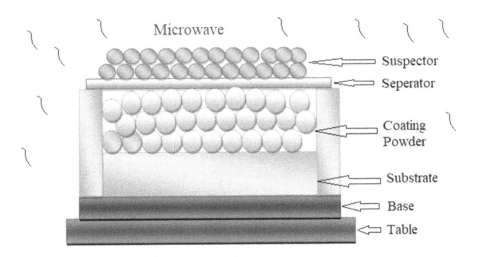

**FIGURE 2.7**  Microwave Cladding Setup.

overcome this problem, charcoal is used as a suspector material, which interacts with microwaves at atmospheric temperature and rapidly heats up. This, in turn, increases the temperature of the clad powders making these interact with microwaves. Pure graphite is used as a separator between the suspector and the clad materials in order to avoid contaminations. Experimental trials are carried out in a microwave oven with a frequency of 2.45 GHz at 900 W. The time of the cladding process varies from 180 seconds to 420 seconds, based on the thickness of the cladding. The workpiece is cooled in the setup for approximately 600 seconds.

### 2.5.3 PROCESS PARAMETERS

The process parameter that affects the microwave cladding is depicted in Figure 2.8. Microwave cladding is suitable for cladding mainly austenitic types of steel that have different grades, copper, aluminium with reinforcements like nickel, tungsten and silicon carbide. Composites cladding on austenitic steels by microwave cladding using nickel and tungsten as clad materials revealed that the cladding exhibits a

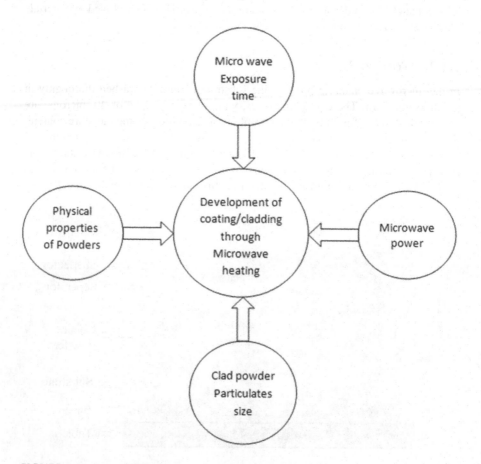

**FIGURE 2.8** Process Parameters Affecting Microwave Cladding.

uniform and dense structure resulting in higher microhardness as well as greater wear resistance (Gupta and Sharma 2011).

Mutual diffusion of elements with the substrate improves the bonding strength characterised by partial dilution. The microwave clads are free from cracks due to solidification and high temperature initiates the formation of carbides. The wear resistance of clad composites by microwave cladding was greater than the substrate. Formation of oxide layers during the process may occur, and this should be avoided (Sharma and Gupta 2012). The clad composites fabricated by the microwave cladding process have less porosity and exhibit higher microhardness (Gupta and Sharma 2014).

Sharma and Gupta (2012) investigated the tungsten cladded austenite stainless steel substrate. The patterns produced by cladding are of cellular-like structure, which may be due to volumetric heating. The hardness of the composite increases by approximately 100% than the base substrate with good deformation resistance. The clad materials exhibit better tenacity and do not peel off from the base metal at average load during the 3-point bend test. Some significant advantages of this technique include capacity for rapid heating and volumetric heating, variety of production, ecological nature and energy and cost-effectiveness.

## 2.6 ROTARY SWAGING

### 2.6.1 INTRODUCTION

Rotary swaging is a precision forming process usually used for the production of tubes, bars and other cylindrical components (Kocich et al. 2017; Kocich et al. 2018). This type of forging process belongs to the net-shape-forming process where the net shape is obtained with minimal process and cutting. Rotary swaging is a forging process used for producing solid cylindrical, hollow and solid parts where the axis can be elongated. Rotary swaging is suitable for manufacturing various parts in aerospace, automotive and defence industries like hollow steering columns, guide shafts, fasteners etc. Rotary swaging increases the durability of the components compared to that of the other processes. Swaging is normally a cold working process, but in some cases, it can be used in a hot working condition. The rotary swaging process involves two categories: (i) the extrusion of the workpiece, forcing it into the die and in order to reduce the diameter of the workpiece and (ii) using two or more dies to hammer the workpiece to be able to lessen the diameter.

### 2.6.2 WORKING PRINCIPLE

The rotary swaging machine is comprised of two to six, or even eight, dies depending upon the application for radial forming. A single die normally consists of four die segments as shown in Figure 2.9. The forming dies are placed concentric around the workpiece in the swaging machine. The swaging die connected to the spindle generally rotates around the workpiece and, in some cases, the workpiece instead rotates around the swaging dies.

In case of non-circular forming, both the workpiece and die will be arrested from rotational movement. The workpiece moves in the forward and backward direction

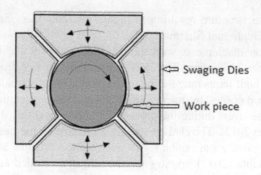

**FIGURE 2.9** Working Principle of Rotary Swaging.

along the axis. The formation in rotary swaging takes place in multiple small processing stages; hence, this is referred to as incremental forming process. The swaging die utilises high frequency, which are short-stroke radial movements of around 10,000 strokes per minute and at a radial displacement of 0.2 mm per stroke. A relative motion between the workpiece and die takes place in order to avoid the formation of longitudinal burrs at the gap of the dies. The major advantage of the incremental forming process is the homogenous material that achieves high metal qualities with less friction in a single processing step with uniform distribution of materials. The major benefits of the rotary swaging process are net-shape production, significant weight savings (30%–50%), high product quality, high forming ratios, no restriction to materials, cold and hot forming, environmental acceptability, versatility, short production times and high-efficiency rates compared to other conventional methods. Other than the standard swaging machine, butt swaging machines, where a set of wedges closes the swaging dies onto the workpiece, are also available.

### 2.6.3 PROCESS PARAMETERS

Composite materials can be fabricated by methods like severe plastic deformation and the conventional forming process. Plastic deformation is one of the techniques used for fabricating clad composites both at cold and hot conditions. There are different forming processes like extrusion, drawing, equal channel angular pressing, accumulative roll bonding, roll bonding, equal channel angular pressing etc. Most of the techniques are unsuitable for fabricating axisymmetric components with a complex profile. Rotary swaging is a method that can be used for producing long products in the form of wires and rods. This technique can also be employed to create clad composite wires made of materials like wires and rods with an excellent surface finish.

Rotary swaging can be used for fabricating clad composites like Al/Cu (Figure 2.10) to improve the mechanical properties, electric resistivity and enhanced structures by means of some thermomechanical treatments. The working temperatures can be varied from 25°C and 250°C. A 30 mm aluminium rod

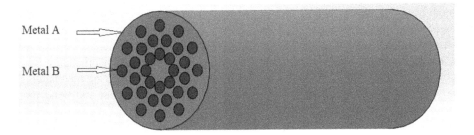

Metal A

Metal B

**FIGURE 2.10**  Clad Composite Specimen before Swaging.

containing 3 mm wire inserts can be rotary swaged to diameters up to 5 mm in six swaging passes. A nonlinear variation in hardness of the composites is induced due to the imposed strain during rotary swaging. Heat treatment processes, such as annealing, after swaging influenced the structural changes like the formation of intermetallics on the clad composite layer. The thermal conductivity and electric conductivity of the clad Al/Cu composites were higher than that of the swaged pure aluminium workpieces. The tensile strength of the composites swaged reached a maximum of 500 MPa and the variation in mechanical property is determined by plasticity. Microhardness distribution can be attained while carrying swaging at higher temperatures (Kocich et al. 2017).

Cu/Al clad composites in the range of 81:19 can be manufactured with progressive grain refinement, utility property, better mechanical properties and good surface accuracy. The major factors influencing the properties are imposed strain and the swaging temperature. Formation of intermetallics, hardening, dislocations density, texture development and recrystallisation at the interfaces reduces the electric conductivity of the composites. Deformation structure characteristics like precipitation, work hardening lead to steady-state reduction ratios (Kocich et al. 2018). Clad composites swaged at higher temperatures heightened the deformation tendency of wires as compared to swaging done at lower temperatures. The plastic flow of the material decreases at higher temperatures (Kocich et al. 2017). A modified version of the rotary swaging process called multi-billet rotary swaging can be used to fabricate bimetal clad composites and even tubes, and trilaminar metal tubes can be made of materials like Ti/Pb/Cu. The deformation trends of the various metals are stable and vary from metal to metal. Multi-billet rotary swaging produces materials with a good interface transition zone that influences a higher bonding strength. Finally, this technique can also be used to fabricate materials like plastics (Wang and Han 2015).

## 2.7  PLASMA SURFACE CLADDING

### 2.7.1  INTRODUCTION

Mechanical parts utilised in chemical and oil fields are often subject to abrasion and corrosion, thus damaging the surface of the components. The surface modifications process has become an important topic of interest. Conventional surface modification methods such as spraying and electroplating have many demerits such as poor bonding of coating and thin density coating, among other things. The shortcomings can be overthrown by using various cladding processes. Plasma arc cladding possesses many favourable characteristics, such as high efficiency, low cost and good fusion properties. The plasma beam is used for melting the cladding powder and the substrate. This type of cladding process can coat and bond metallurgically with bulk material.

### 2.7.2 WORKING PRINCIPLE

Initially, the substrate is machined for a uniform surface, and then acetone is utilised to sanitise the samples for the removal of oil or grease. The powder used for cladding is preheated to a temperature of 120°C for two hours in order to release the moisture content. Figure 2.11 presents the working principle of plasma cladding technology. The heat produced by the plasma arc creates a thin metal pool on the substrate. Simultaneously, cladding powders are fired into the melt pool through the powder feeder. The shielding gas (Ar) is used to protect the melt pool zone from contamination caused by the harmful gases in the atmosphere. Water is utilised to reduce the nozzle temperature. Once the cladding is completed, low-temperature air is used for solidification.

### 2.7.3 PROCESS PARAMETERS

Plasma surface cladding is influenced by parameters like current, voltage, scanning velocity, powder feeding rate, feeding gas flow, protective gas flow, plasma gas flow, plasma arc length etc.

Ferrous-based composite plasma is cladded with $CeO_2$ and $La_2O_3$. The results reveal that the clad powder addition refined the microstructure of the coating. The microhardness was considerably increased with respect to the base alloy (Zhang et al. 2008).

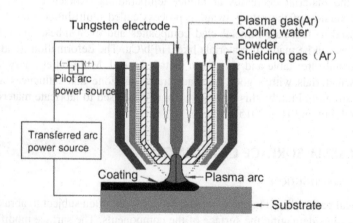

**FIGURE 2.11** Schematic Diagram of Plasma Arc Setup.

The mixture of Ni-coated WC coatings and Nickel-based alloys were coated on a steel substrate by a high-temperature plasma jet exhibited good bonding with very low porosity between the powder and substrate. Electrochemical behaviour, microhardness and cladding depend on the mixture of Ni-coated WC powder. The corrosion mechanisms of clad composites are influenced by the microgalvance corrosion that exists in between the phases of coatings (Guozhi et al. 2010).

The absence of porosity was observed in nickel-coated tungsten carbide in the absence of cracks in Ni60A alloys. An increase in clad materials also heightened the microhardness of the composites (Guojian et al. 2006).

In an interesting study, carbon steel was coated with ferrous alloy using the plasma cladding technique. An abrasive test of the cladded composites shows better wear resistance when compared to the quenched mild steel, thus contributing better microstructures. The carbide and boride were homogeneously dispersed in the dendritic austenite. The study also reported the plasma cladding method has greater efficiency and improved cladding quality (Zhang et al. 2007).

## 2.8  CONCLUSION

A comprehensive review on the various cladding processes, namely: roll bonding, accumulative rolling bonding, friction surface cladding, laser cladding, microwave cladding, rotary swaging and plasma surface cladding were discussed in this study. A detailed discussion on the various cladding processes with respect to their operating mechanism, process parameters, advantages, limitations and applications was explained. Clad composites are utilised in processing materials like stainless steel, nickel alloys, copper alloys, aluminium alloys etc. Cladding also involves the coating of materials using hard particles such as SiC, $B_4C$ and $Y_2O_3$ NiTi, $TiB_2$, $Cr_7C3$, CoTi, $CoTi_2$, $Cr_7C_3$, TiC, TiB and $Ti_5Si_3$ etc. The ability of the cladding process in fabricating components using different materials can be a proficient process in manufacturing industries.

## REFERENCES

Akramifard, H. R., H. Mirzadeh and M. H. Parsa. 2014. "Cladding of aluminum on AISI 304L stainless steel by cold roll bonding: mechanism, microstructure, and mechanical properties". *Materials Science and Engineering: A*, 613, Elsevier: 232–239.

Altenbach, Holm, Johannes Altenbach and Wolfgang Kissing. 2018. "Classification of composite materials". *Mechanics of Composite Structural Elements*. Singapore: Springer.

Badheka, K. and V. Badheka. 2017. "Friction surfacing of aluminium on steel: an experimental approach". *Materials Today: Proceedings*, 4(9), Elsevier: 9937–9941.

Da Silva, L., M. El-Sharif, C. Chisholm and S. Laidlaw. 2014. "A review of the cold roll bonding of AlSn alloy/steel bimetal strips". Paper presented at 23rd International Conference on Metallurgy and Materials, Brno, Czech Republic.

Dhib, Z., N. Guermazi, M. Gaspérini and N. Haddar, 2016. "Cladding of low-carbon steel to austenitic stainless steel by hot-roll bonding: Microstructure and mechanical properties before and after welding". *Materials Science and Engineering: A*, 656, Elsevier: 130–141.

Doubenskaia, M., A. K. Gilmutdinov and K. Y. Nagulin. 2015. "Laser cladding of metal matrix composites reinforced by cermet inclusions for dry friction applications at ambient and elevated temperatures". *Surface and Coatings Technology*, 276, Elsevier: 696–703.

Gandra, J., H. Krohn, R. M. Miranda, P. Vilaça, L. Quintino and J. F. Dos Santos. 2014. "Friction surfacing—A review". *Journal of Materials Processing Technology*, 214(5), Elsevier: 1062–1093.

Ghalehbandi, Seyed Mahmoud, Massoud Malaki and Manoj Gupta. 2019. "Accumulative roll bonding - A review". *Applied Sciences*, 9, MDPI: 1–32.

Guojian, Xu, Muneharu Kutsuna, Zhongjie Liu and Hong Zhang. 2006. "Characteristics of Ni-based coating layer formed by laser and plasma cladding processes". *Materials Science and Engineering A*, 417, Elsevier: 63–72.

Guozhi, X., S. Xiaolong, Z. Dongjie, W. Yuping and L. Pinghua. 2010."Microstructure and corrosion properties of thick WC composite coating formed by plasma cladding". *Applied Surface Science*, 256(21), Elsevier: 6354–6358.

Gupta, D. and A. K. Sharma. 2011. "Investigation on sliding wear performance of WC10Co2Ni cladding developed through microwave irradiation". *Wear*, 271(9–10), Elsevier: 1642–1650.

Gupta, D. and A. K. Sharma. 2014. "Microwave cladding: A new approach in surface engineering". *Journal of Manufacturing Processes*, 16(2), Elsevier: 176–182.

Ha, Jong Su and Sun Ig Hong. 2013. "Design of high strength Cu alloy interlayer for mechanical bonding Ti to steel and characterization of their tri-layered clad". *Materials and Design*, 51, Elsevier: 293–299.

Han, T., M. Xiao, J. Zhang, X. Feng and Y. Shen. 2018. "Laser cladding composite coating on mild steel using Ni–Cr–Ti–B4C powder". *Surface Engineering*, 36(12), Taylor & Francis: 1–7.

Jamaati, R. and M. R. Toroghinejad. 2010. "Investigation of the parameters of the cold roll bonding (CRB) process". *Materials Science and Engineering: A*, 527(9), Elsevier: 2320–2326.

Jamaati, R. and M. R. Toroghinejad. 2011. "Cold roll bonding bond strengths: Review". *Materials Science and Technology*, 27(7), Taylor and Francis:1101–1108.

Janicki, D. 2017. "Laser cladding of Inconel 625-based composite coatings reinforced by porous chromium carbide particles". *Optics & Laser Technology*, 94, Elsevier: 6–14.

Kocich, R., L. Kuncicka, P. Král and P. Strunz. 2018. "Characterization of innovative rotary swaged Cu-Al clad composite wire conductors". *Materials & Design*, 160, Elsevier: 828–835.

Kocich, R., L. Kuncicka, A. Machackova and M. Šofer. 2017. "Improvement of mechanical and electrical properties of rotary swaged Al-Cu clad composites". *Materials & Design*, 123, Elsevier: 137–146.

Kumar, B. V., G. M. Reddy and T. Mohandas. 2015. "Influence of process parameters on physical dimensions of AA6063 aluminium alloy coating on mild steel in friction surfacing". *Defence Technology*, 11(3), Elsevier: 275–281.

Li, L., K. Nagai and F. Yin. 2008. "Topical review: Progress in cold roll bonding of metals". *Science and Technology of Advanced Materials*, 9(2), IOP: 1–12.

Liu, S., T. C. Bor, A. A. Van der Stelt, H. J. M. Geijselaers, C. Kwakernaak, A. M. Kooijman, J. M. C. Mol, R. Akkerman and A. H. Van Den Boogaard. 2016. "Friction surface cladding: An exploratory study of a new solid state cladding process". *Journal of Materials Processing Technology*, 229, Elsevier: 769–784.

Mozaffari, A., H. D. Manesh and K. Janghorban. 2010. "Evaluation of mechanical properties and structure of multilayered Al/Ni composites produced by accumulative roll bonding (ARB) process". *Journal of Alloys and Compounds*, 489(1), Elsevier: 103–109.

Paydas, H., A. Mertens, R. Carrus, J. Lecomte-Beckers and J. T. Tchuindjang. 2015. "Laser cladding as repair technology for Ti–6Al–4V alloy: Influence of building strategy on microstructure and hardness". *Materials & Design*, 85, Elsevier: 497–510.

Ravi Kumar, K., T. Pridhar and V. S. Sree Balaji. 2018. "Mechanical properties and characterization of zirconium oxide ($ZrO_2$) and coconut shell ash (CSA) reinforced aluminium (Al 6082) matrix hybrid composite". *Journal of Alloys and Compounds*, 765, Elseiver: 171–179.

Ray, A., K. S. Arora, S. Lester and M. Shome. 2014. "Laser cladding of continuous caster lateral rolls: Microstructure, wear and corrosion characterisation and on-field performance". *Journal of Materials Processing Technology*,214, Elsevier: 1566–1575.

Sardar, S., A. Mandal, S. K. Pal and S. B. Singh. 2015. "Solid-state joining by roll bonding and accumulative roll bonding". In *Advances in Material Forming and Joining*, Narayanan, R. Ganesh, and Uday Shanker Dixit. New Delhi, India: Springer, 351–377.

Saw, K., S. Shankar, S. Chattopadhyaya and P. Vilaca. 2018. "Microstructure evaluation of different materials after friction surfacing-A review". *Materials Today: Proceedings*, 5(11), Elsevier: 24094–24103.

Sharma, A. K. and D. Gupta. 2012. "On microstructure and flexural strength of metalceramic composite cladding developed through microwave heating". *Applied Surface Science*, 258(15), Elsevier: 5583–5592.

Shepeleva, L., B. Medres, W. D. Kaplan, M. Bamberger and A. Weisheit. 2000. "Laser cladding of turbine blades". *Surface and Coatings Technology*, 125(1–3), Elseivier: 45–48.

Venkatesh, L., T. V. Arjunan and K. Ravi Kumar, 2019. "Microstructural characteristics and mechanical behavior of aluminium hybrid composites reinforced with groundnut shell ash and $B_4C$". *Journal of the Brazilian Society of Mechanical Sciences and Engineering*, 41(295), Springer: 1–13.

Wang, H. F. and J. T. Han. 2015. "Fabrication of laminated-metal composite tubes by multi-billet rotary swaging technique". *The International Journal of Advanced Manufacturing Technology*, 76(1–4), Springer: 713–719.

Wang, Yibo, Shusen Zhao, Gao Wenyan, Zhou Chunyang, Falan Liu and Xuechun Lin. 2014. "Microstructure and properties of laser cladding FeCrBSi composite powder coatings with higher Cr content". *Journal of Materials Processing Technology*, 214, Elsevier: 899–905.

Weng, F., H. Yu, C. Chen and J. Dai. 2015. "Microstructures and wear properties of laser cladding Co-based composite coatings on Ti–6Al–4V". *Materials & Design*, 80, Elsevier: 174–181.

Zhang, L., D. Sun and H. Yu. 2007. "Characteristics of plasma cladding Fe-based alloy coatings with rare earth metal elements". *Materials Science and Engineering: A*, 452, Elsevier: 619–624.

Zhang, L., D. Sun and H. Yu, 2008. "Effect of niobium on the microstructure and wear resistance of iron-based alloy coating produced by plasma cladding". *Materials Science and Engineering: A*, 490(1–2), Elsevier: 57–61.

# 3 A Brief Review on Green Manufacturing

## P. M. Gopal, V. Kavimani, and Senol Bayraktar

## CONTENTS

## 3.1 INTRODUCTION

The industrial and manufacturing sector has become quite advanced via technological development in terms of fulfilling a wide range of accelerated customer requirements. On the other hand, technological advancements for the manufacturing of quality products are creating negative impacts such as environmental pollution. Broadly, environment degradation in terms of natural resource deployment and pollution, is a major global issue. The International Organisation for Standardisation (ISO), a governing body that proposes quality-related management systems, has developed measures for an environment management system with an aim to reduce damage caused by the industries. Alongside this, in order to address this issue and achieve the aforementioned standards, a new strategy and set of techniques are necessitated: this is green manufacturing for sustainable production and development (Tan et al. 2002).

There is a steady increase found in the price of energy and resources that are utilised for manufacturing, as a result of increased demand and inadequate supply. In addition, it is rather difficult to forecast the price trend, and manufacturers are producing the same products within large energy and resource price ranges. A balance can be maintained by minimising the consumption of resources and improving the correlation among manufacturing systems (Diaz-Elsayed et al. 2013).

Since the growth in industrialisation is unavoidable, there is a need to mitigate the global environmental concerns in which the manufacturing sector should diminish waste generation as well as plan for efficient energy and resource utilisation. In this century, the temperature of the universe is estimated to increase by 0.6°C even after the complete shutdown of factories, power plants and means of transportation. To avoid such an event, sustainable or green manufacturing practices must be exploited (Figure 3.1). Green manufacturing is implemented in one of the three ways:

1. green energy usage
2. green product development and selling
3. employing green processes in business operations.

A recent global survey reveals that nearly 92% of surveyed industries are engaged in green initiatives. Various reasons cause different industries to adhere to green manufacturing practices, such as being compelled or staying competitive. For example, power generation sectors such as thermal power stations follow green techniques as a result of regulatory compulsion, whereas the auto manufacturers utilise this prospect to create pollution-free electric and efficient hybrid vehicles to meet up increasingly severe emission policies. On the other hand, retail sectors engage in order to remain in global competition.

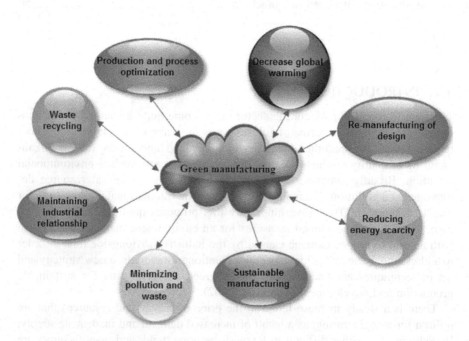

**FIGURE 3.1** Scope of Green Manufacturing.

The aim and impact of utilising green initiatives vary from industry to industry, akin to these initiatives is the impact on dropping $CO_2$ emissions in the power sector, followed by the transportation and then the industrial sector. Contrarily, there is a considerable increase in the trend of adopting green products and habits by consumers themselves. It was concluded from a survey that more than 50% of people in both developed and developing countries gave preference to green products, mainly in food and consumer durables. Furthermore, most of them were willing or ready to pay a higher amount for green products. However, it was also revealed from the survey that there is also a colossal gap in consumer awareness about green products that companies must associate with.

## 3.2 IMPLEMENTATION OF GREEN MANUFACTURING

It is mandatory to discuss certain small, individual initiatives for the successful implementation of green manufacturing. The following three-step framework can be adopted for this reason:

- The core part of the business strategy should be planned for green.
- Initiatives should be taken from corner to corner of the value chain by transferring en route for green energy, green products and green processes.
- The benefits of green initiatives have to be communicated and promoted to all stakeholders.

### 3.2.1 TRANSFORMATION TO GREEN MANUFACTURING

Researchers have exposed that green manufacturing spotlights primarily on environmental pollution, energy effectiveness in manufacturing and waste management (Rehman and Shrivastava 2013). Three areas that the manufacturing companies can deal with in order to transform their ventures to become green are:

I. *Green energy*

The first step involves the production and use of cleaner energy.

- independence of industry for energy
- arranging renewable energy sources like natural gas, wind, solar and biomass.
- achieving higher energy effectiveness during operations.

II. *Green products*

The second step is to focus on developing green products.

- manufacturing 'recycled', 'low carbon footprint', 'organic' and 'natural' products

- manufacturing green products and effectively marketing them to the pertinent consumers
- even it can often lead to higher costs; manufacturers can balance this by developing additional volumes and with a premium price

### III. *Green processes in business operations*

Implementing green processes across the operations is the third aspect of green manufacturing that improves operational efficiency and lowers costs.

- efficient use of key resources
- minimising waste creation through lean procedures
- playing down the carbon footprint and water conservation

Green manufacturing help manufacturers to make a profit in a sustainable way and increases environmental safety. It includes various techniques that focus on material waste minimisation, recycling, a changeover to less hazardous substitutes, internal usage of wastes and remanufacturing (Worrell and Price 2001). In green manufacturing, importance is set to reduce the environmental concerns by reducing, reusing and re-manufacturing that lead to source cutback and resource utilisation minimisation (Shang, Lu, and Li 2010). The main essence of green manufacturing is on transformation toward environmentally sound technologies in conjunction with monetary services (Shrivastava and Shrivastava 2017; Varma, Wadhwa, and Deshmukh 2006).

## 3.3   GREEN TOOLS AND PRINCIPLES

### 3.3.1   GREEN PRINCIPLES

It is necessary to set up a suitable approach to be able to realise green manufacturing in order to attain a healthy environment. Anastas and Zimmerman (2003) proposed certain green manufacturing principles for green process design. They also emphasised that there should be an evaluation for energy input and output and the nonhazardous nature of materials and lifecycle as well. Those green principles were further simplified by Helu and Dornfeld (2013), and the five-principle framework was proposed. Table 3.1 summarises some of the green principles proposed by researchers.

Furthermore, various authors suggest a number of factors that have to be considered in green manufacturing. The main design elements for the product lifecycle that have to be integrated with the green manufacturing system, as proposed by Lele (2009), are design, acquisition, production, packing and supply, product end of life and remanufacturing. These proposed product lifecycle element details are further expanded and explained with a more detailed framework by Vila et al. (2015) by way of concentrating on the technique, tools and knowledge in lifecycle design. It was also categorised into three key divisions under the title of design, manufacturing and services. It is more centred on implementing a green approach on pre-manufacturing, such as design during manufacturing and post-manufacturing like service. The framework proposed by both Vila et al. (2015) and Lele (2009) mainly

**TABLE 3.1**

**Green Principles (Helu and Dornfeld 2013; Anastas and Zimmerman 2003)**

| Sl No. | Description of Principle |
|--------|--------------------------|
| 1 | need for a comprehensive system approach to mitigate the negative impacts of manufacturing processes |
| 2 | improvement of recycle, reuse and remanufacturing practices |
| 3 | energy and resource efficiency |
| 4 | equal focus on the long term as well as short term impacts |
| 5 | regular and frequent feedback and supervision |
| 6 | innovations for new resources, materials and processes |

focuses on the management of resources, control of production, fabrication, assembly and package storage.

Newer technology with superior ecological performance, green energy utilisation and design for sustainable development are the aspects that have to be taken as measures for green manufacturing. It is also a well-known fact that the financial, societal and ecological impact of any product can be identified by nearly 80% when the product/process design is completed (Kim and Kara 2012). Therefore, it is crucial to have a product lifecycle assessment (LCA) at the preliminary stage of design (Ma et al. 2012). The significance of lifecycle assessment in assessing the product that causes ecological contamination is further discussed by Paul, Bhole and Chaudhari (2014).

Lifecycle assessment is done to find the probable ecological influence of any product throughout its lifetime, from unprocessed material, manufacturing, application to its disposal. Accordingly, it is rather important to include or introduce the green approach in the design stage itself in the name of design for the environment (DfE). According to Eibel and Joanneum (2014), the ecological impact of any product all the way through its lifecycle is identified and designed accordingly in the design of the environment approach. Minimising the environmental emission or waste is the relatively close relationship between design for the environment and lifecycle assessment. Energy utilisation has a noteworthy part in product manufacturing excluding DfE and LCA.

### 3.3.2 GREEN ENERGY

Nowadays, the word 'green technology' is not unknown even for a small-scale industrialist. According to Ishak, Jamaludin and Abu (2017), the application of science to the ecosystem in the view of controlling the negative effect of activities done by humans is referred to as green technology. In the recent past, the green aspect of a factory is enhanced with energy–efficient tools like photovoltaic, cogeneration, biogas, etc., and this gained wide attention. However, these green tools have only minimal share in the market even though these reached a maturity level in the technological aspects. Technology level, system perspective, paradigm shift and system knowledge level are the areas where the changes are suggested for green manufacturing (Yong 2014). Screening techniques can be utilised to assess newer technologies from DfE and LCA of the product of a process that can remove or decrease effluence (Reich-Weiser, Vijayaraghavan and Dornfeld 2010). Green technology in an industrial unit is able to bring an affirmative impact in both management of energy and performance optimisation. For example, it is possible to minimise the cost of operations exclusive of compromise in exhaust emission in gas turbine cogeneration systems. The idea of including ecological distress into material planning was proposed by Melnyk et al. (2001) who included these concepts with traditional material requirement planning system (MRP). This proposed system is useful for superior understanding in both quantitative and economic aspects of the possibilities of an operation's waste stream generation. On the other hand, in the view of evaluating the manufacturing environmental hazards, a network analytic model is also suggested (Hui, He, and Dang 2002). This proposal concentrates on examining the causes created by various wastes coming from manufacturing. A bottom-up analysis mode-based ecological value system examination tool was developed by Krishnan, Raoux and Dornfeld (2004) in order to evaluate the ecological performance that facilitates the flexibility in analysis requisite by the user.

The benefit of green technology is not only minimising environmental pollution, but it also makes condensed operation costs possible. Cogeneration or trigeneration, such as well-managed green technologies, can increase the energy generation efficiency of the facilities. A systematic approach was proposed by Andiappan and Ng (2016) in selecting the trigeneration system by which improvement in the energy performance of the industry players can be achieved.

### 3.3.3 INDUSTRIAL SYMBIOSIS

Berkel (2010) suggested that industrial symbiosis permits a factory to replace resources with exhausted materials, such as wastes. Conversion, replacement and avoidance are the steps involved in the substitution procedure, and in the way towards green manufacturing. The first step, named as conversion, refers to the identification and gathering of symbiosis partner's waste that is possible to utilise as a feedstock or energy resource and with the possibility of being transported to the required location. Since the waste materials are used as substitute feedstock or energy resources, then it is necessary to restructure or modify the process according to the requirement, which is precisely the second step called substitution/replacement. The third step, avoidance, completely concentrates on minimisation or elimination of waste generated while processing that which is capable to harm the environment.

## 3.4 DRIVERS AND BARRIERS FOR GREEN MANUFACTURING

### 3.4.1 GREEN AND LEAN MANUFACTURING

Green manufacturing is somewhat identical with lean manufacturing in the face of initiatives for waste reduction. Some researchers such as Hines (2009) came up with different approaches, namely the L&G approach, in order to close the gap between these manufacturing technologies and to align green waste with lean waste. Similarly, in the view of performance improvement in operations, certain correlation is proposed between lean waste and green waste by Verrier, Rose and Caillaud (2016). It was also advised to incorporate both lean and green wastes together as singular waste for additional enhancement. Greenhouse gases, eutrophication, resource utilisation in excess, power usage in excess, pollution, rubbish, excessive water usage and poor health and safety are termed as green wastes.

### 3.4.2 EXTERNAL DRIVING FORCES

Legislation, peer pressure, corporate likeness and incentives for tax are the key driving forces that will come from the external side of the industry. In terms of supremacy, legality and pressure, the influence on the organisation by the stakeholders such as supplier, consumer, government bureau, non-governmental organisation and investors are significant en routes for green manufacturing (Álvarez-Gil et al. 2007).

### 3.4.3 GOVERNMENT REGULATIONS

In the process of encouraging and advocating the manufacturers in the direction of accomplishing green manufacturing, government regulations are playing an imperative role (Zhu, Sarkis, and Geng 2005). Several governments are also offering some tax incentives such as investment tax allowances to the industry for practicing green activities. The companies will propel the green implementations and gain confidence in green manufacturing if the support is given by the government.

The Japanese government shows strong commitment and takes great initiatives towards the green approach. In the view of the increasing possibility of low-carbon technology and system implementation in certain emerging countries, the Joint Crediting Mechanism (JCM) was facilitated by Japan. Greenhouse gas emission minimisation or removal is the benefit that Japan will get as a return for offering JCM and helping to achieve emission reduction targets.

To realise green in the company, there are many tools and techniques that are available to assist. For instance, continuous following of green policies is mandatory in order to retain the official recognition for an ISO-certified corporation. It was also discovered from the analysis that there are nearly 40% of the ISO 14,001 certified industries expected to evaluate the environmental performance of their suppliers, whereas the industries asking their suppliers to carry out definite environmental practices are at approximately 50% (Maloni and Brown 2006).

In the present scenario, awareness and education about climate change for clients are increasing. In order to implement green manufacturing in industry, pressure from the client also plays a vital role in enforcing green technologies (Carter and Jennings 2002). Regular clients also put enormous pressure on the manufactures and demand green initiatives (Hall 2006). This demand and pressure from the clients also depend on the pressure from competitors. Companies are quite aware that the ecological associated corporate societal responsibility will improve their corporate image.

### 3.4.4 Barriers

There is a discussion regarding green technology with a number of companies from different sectors, such as pharmaceutical companies, rubber processing facilities and palm oil refineries. These companies that were considered for discussion are the active production industries and major economic players of Southeast Asia. The conversation made it clear that there are small numbers of vital facets that are likely to set back the execution of green approaches in the factory.

- The main factor that hinders the implementation of green technology is the inability of the management to find the prospective or possibility to enhance the factory operator's competence.
- Delay in green implementations is also due to the deficiency in expertise present in the industrial unit.
- The statement given by some of the engineers is that there is the only possibility for nil or low-priced enhancement in the industry.
- Industry management does not have the employee with the capability of leading the project and unfamiliarity with the technology.
- Another aspect that acts as a key player in convincing management for green execution is investment payback.
- Generally, the main concern for the manufacturing industry is green technology and financial scheme.
- Another major problem is that the companies are only concentrating on a few, smaller energy-efficient components because only 2%–5% of improvement is possible.
- It was also discovered that there are only some particular departments in the factory that are actively involved in green technology initiation and implementation.

## 3.5 APPLICATIONS OF GREEN MANUFACTURING

Green manufacturing has applications in the following areas and more:

- minimising waste generation besides energy reduction
- recycle, reuse, and remanufacturing materials and products
- from design to production and logistics
- green raw material, chemical, tools, consumables and processes.
- making production energy and resource-efficient.

## 3.6 SUMMARY

A brief review of various aspects of green manufacturing has been presented in this chapter. Green manufacturing mainly stresses the minimisation of resources, energy and environmental emission of an industry. Nowadays, industrial sectors all over the world are focusing on green manufacturing implementation. The government and agencies are busy making regulations and providing incentives on following green practices. Updated technology with superior ecological performance, green energy utilisation and design for sustainable development are the key components that have to be taken as measures in order to overcome problems that are related to environmental degradation. Technology level, system perspective, paradigm shift, employee training and system knowledge level are the areas where these changes have to be done for green manufacturing. Legislations and regulations, peer pressure, corporate likeness and incentives for green manufacturing are the key drivers simple enforcement of green strategies and techniques.

## REFERENCES

Álvarez-Gil, Ma Pascual, José Berrone, F. Javier Husillos, and Nora Lado. 2007. "Reverse Logistics, Stakeholders' Influence, Organizational Slack, and Managers' Posture." *Journal of Business Research* 60 (5). Elsevier: 463–473.

Anastas, Paul T., and Julie B. Zimmerman. 2003. *Peer Reviewed: Design through the 12 Principles of Green Engineering* USA: ACS Publications.

Andiappan, Viknesh, and Denny K. S. Ng. 2016. "Synthesis of Tri-Generation Systems: Technology Selection, Sizing and Redundancy Allocation Based on Operational Strategy." *Computers & Chemical Engineering* 91. Elsevier: 380–391.

Berkel, René Van. 2010. "Quantifying Sustainability Benefits of Industrial Symbioses." *Journal of Industrial Ecology* 14 (3). Wiley Online Library: 371–373.

Carter, Craig R., and Marianne M. Jennings. 2002. "Social Responsibility and Supply Chain Relationships." *Transportation Research Part E: Logistics and Transportation Review* 38 (1). Elsevier: 37–52.

Diaz-Elsayed, Nancy, Annabel Jondral, Sebastian Greinacher, David Dornfeld, and Gisela Lanza. 2013. "Assessment of Lean and Green Strategies by Simulation of Manufacturing Systems in Discrete Production Environments." *CIRP Annals* 62 (1). Elsevier: 475–478.

Eibel, David, and F. H. Joanneum. 2014. "Green Manufacturing An Essential Success Factor in a Globalized World." *Austrian Marshall Plan Foundation, FH JOANEUM University of Applied Sciences Styria–Austria.*

Hall, Jeremy. 2006. "Environmental Supply Chain Innovation." In *Greening the Supply Chain*, 233–249. UK: Springer.

Helu, Moneer, and David Dornfeld. 2013. "Principles of Green Manufacturing." In *Green Manufacturing*, 107–115. Springer.

Hines, Peter. 2009. "Lean and Green." *Source Magazine the Home of Lean Thinking* 3.

Hui, I. K., L. He, and C. Dang. 2002. "Environmental Impact Assessment in an Uncertain Environment." *International Journal of Production Research* 40 (2). Taylor & Francis: 375–388.

Ishak, Ismaniza, Roslan Jamaludin, and Noor Hidayah Abu. 2017. "Green Technology Concept and Implementataion: A Brief Review of Current Development." *Advanced Science Letters* 23 (9). American Scientific Publishers: 8558–8561.

Kim, Seung Jin, and Sami Kara. 2012. "Impact of Technology on Product Life Cycle Design: Functional and Environmental Perspective." In *Leveraging Technology for a Sustainable World*, 191–196. Springer.

Krishnan, Nikhil, Sebastien Raoux, and David Dornfeld. 2004. "Quantifying the Environmental Footprint of Semiconductor Equipment Using the Environmental Value Systems Analysis (EnV-S)." *IEEE Transactions on Semiconductor Manufacturing* 17 (4). IEEE: 554–561.

Lele, Satish. 2009. "Getting Serious about Green Manufacturing." *JO Market Insight Asia Pacific Industrial Technologies Frost and Sullivan.*

Ma, Jian, Fengfu Yin, Zhenyu Liu, and Xiaodong Zhou. 2012. "The Eco-Design and Green Manufacturing of a Refrigerator." *Procedia Environmental Sciences* 16. Elsevier: 522–529.

Maloni, Michael J., and Michael E. Brown. 2006. "Corporate Social Responsibility in the Supply Chain: An Application in the Food Industry." *Journal of Business Ethics* 68 (1). Springer: 35–52.

Melnyk, S. A., R. P. Sroufe, F. L. Montabon, and T. J. Hinds. 2001. "Green MRP: Identifying the Material and Environmental Impacts of Production Schedules." *International Journal of Production Research* 39 (8). Taylor & Francis: 1559–1573.

Paul, I. D., G. P. Bhole, and J. R. Chaudhari. 2014. "A Review on Green Manufacturing: It's Important, Methodology and Its Application." *Procedia Materials Science* 6. Elsevier: 1644–1649.

Rehman, Minhaj A. A., and R. L. Shrivastava. 2013. "Green Manufacturing (GM): Past, Present and Future (a State of Art Review)." *World Review of Science, Technology and Sustainable Development* 10 (1–3). Inderscience Publishers Ltd: 17–55.

Reich-Weiser, Corinne, Athulan Vijayaraghavan, and David Dornfeld. 2010. "Appropriate Use of Green Manufacturing Frameworks." Proceedings of the 17th CIRP Life Cycle Engineering Conference.

Shang, Kuo-Chung, Chin-Shan Lu, and Shaorui Li. 2010. "A Taxonomy of Green Supply Chain Management Capability among Electronics-Related Manufacturing Firms in Taiwan." *Journal of Environmental Management* 91 (5). Elsevier: 1218–1226.

Shrivastava, Sanjeev, and R. L. Shrivastava. 2017. "A Systematic Literature Review on Green Manufacturing Concepts in Cement Industries." *International Journal of Quality & Reliability Management* 34 (1). Emerald Publishing Limited: 68–90.

Tan, X. C., F. Liu, H. J. Cao, and H. Zhang. 2002. "A Decision-Making Framework Model of Cutting Fluid Selection for Green Manufacturing and a Case Study." *Journal of Materials Processing Technology* 129 (1–3). Elsevier: 467–470.

Varma, Siddharth, Subhash Wadhwa, and S. G. Deshmukh. 2006. "Implementing Supply Chain Management in a Firm: Issues and Remedies." *Asia Pacific Journal of Marketing and Logistics* 18(3). Emerald Group Publishing Limited: 223–243.

Verrier, Brunilde, Bertrand Rose, and Emmanuel Caillaud. 2016. "Lean and Green Strategy: The Lean and Green House and Maturity Deployment Model." *Journal of Cleaner Production* 116. Elsevier: 150–156.

Vila, C., José Vicente Abellán Nebot, J. C. Albiñana, and G. Hernández. 2015. "An Approach to Sustainable Product Lifecycle Management (Green PLM)." *Procedia Engineering* 132. Elsevier: 585–592.

Worrell, Ernst, and Lynn Price. 2001. "Policy Scenarios for Energy Efficiency Improvement in Industry." *Energy Policy* 29 (14). Elsevier: 1223–1241.

Yong, L. I. 2014. "Emerging Green Technologies for the Manufacturing Sector," 1–52.

Zhu, Qinghua, Joseph Sarkis, and Yong Geng. 2005. "Green Supply Chain Management in China: Pressures, Practices and Performance." *International Journal of Operations & Production Management* 25 (5). Emerald Group Publishing Limited: 449–468.

# 4 A Review on Additive Manufacturing Technologies

*Jeet Kumar Sahu and Kushagra Tiwari*

## CONTENTS

## 4.1  INTRODUCTION

Additive manufacturing (AM) is poised to bring about significant changes in the designing, manufacturing and distribution of the products. It has achieved prominent academic and industrial interest on account of its ability to fabricate complex geometries using materials that can also be customised. It is an elegant concept of fabrication introduced by the AM in which parts are building by adding materials in the form of multiple layers of thin cross-section according to the information provided by the digital 3D model of the parts.

AM is described as a revolution in product development and manufacturing. This technology has transformed the manufacturing world in multiple dimensions; the rapidness of AM is the key factor, and it is not limited to the time necessary to build the parts. In addition, AM has revolutionised the whole product development cycle with the aid of various new technologies. It has brought significant improvement in manufacturing and benefited a broad application spectrum that includes the automotive industry, construction, healthcare, aerospace and food supply chain (Matta et al., 2018). It can fabricate parts with complex geometry in only a few steps as compared to that in most of the other manufacturing processes wherein parts with complex geometries are fabricated in multiple and iterative stages. Therefore, AM can be considered as a more reliable and precise way to utilise manufacturing time more effectively. It saves time and energy, reduces material wastage and optimises production. For example, the production time for the manufacturing of multi-layer PCB by conventional manufacturing method is reduced by 80% when AM techniques are implemented (Puma-Araujo et al., 2016).

Additive manufacturing is the consequence of developments in computer technology, computer-aided design technology and computer numerically controlled machining. These technologies' collaboration with AM makes it possible to fabricate parts by printing these in three dimensions. Some other associated technologies, such as laser and printing technology, have played an important role in the further development of AM, Ion Gibson et al. (2010, p.17). The historical analysis reveals that in 1980 when a specific generic format capable of manipulating CAD file to drive AM machine was developed by 3D Systems, USA, the AM technology was commercialised.

According to the American Society for Testing and Materials (ASTM), modern AM technologies can be categorised into seven major segments, namely: vat photopolymerization, powder bed fusion, material extrusion, binder jetting, sheet lamination, material jetting and direct energy deposition (Zindani and Kumar, 2019). The main points of differentiation of these segments are in the materials that can be used, the method to be adopted for the creation of subsequent layers and procedure for the bonding of subsequent layers with one another. Figure 4.1 presents all of the seven segments of additive manufacturing.

## 4.2 ADDITIVE MANUFACTURING PROCESSES

Some important additive manufacturing processes that are used in industries are briefly described with the help of sketches.

### 4.2.1 POWDER BED FUSION

Powder bed fusion (PBF) processes are AM techniques that utilise heat to melt powdered material deposited across a thin cross-section so that fusion of material particles can take place. There are four types of PBF-based AM processes, and this division is based on the source of heat energy utilised, environmental condition of the process (e.g. vacuum or inert gas shielding) and materials used. These are

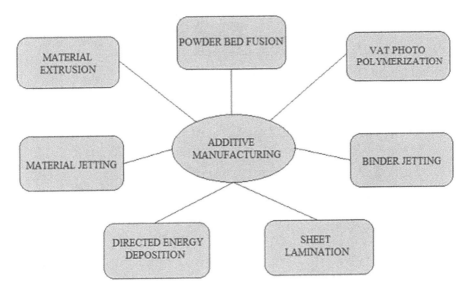

**FIGURE 4.1** Classification of Additive Manufacturing.

selective laser melting (SLM), selective laser sintering (SLS), direct metal laser sintering (DMLS) and electron beam melting (EBM).

In the **selective laser melting process**, a laser beam is used to melt the powdered materials. This process is carried out in an inert gas shielding inside the chamber to diminish in-situ oxidation. In this procedure, process environmental conditions are easy to facilitate, but heat loss due to convection makes the heating process difficult. Consequently, the chances of thermal distortion in products are higher. Ti64 alloy fabricated by SLM shows better mechanical properties as compared to other sintering processes (Q. Yan et al., 2018). The material used in SLS are ceramic (Sudarev et al., 2018), polymers (Stichel et al., 2018), plastic and composites of metals and ceramics (Halloran et al., 2011). In Figure 4.2, the working principle of SLM process is illustrated.

SLM produces good surface finish and high part accuracy. However, the necessity to maintain an inert gas environment results in high convective heat loss and an increase in the distortion of the parts.

**The selective laser sintering (SLS)** process is another PBF-based AM technique in which sintering of powdered material is achieved by using a high-powered carbon-dioxide laser. This laser selectively sinters the powder particles at the particular location, as per information provided by 3D model data. After the formation of one layer, a piston lowers the bed by a length equal to layer thickness, and the formation of another layer over this takes place. The mechanical properties of parts built by SLS are greatly affected by the heating process and can be improved by achieving uniform heating through an automated laser control system (Phillips et al., 2018).

**The electron beam melting** process is based on a powder bed fusion system that makes use of an electron beam as the heat energy source. This process

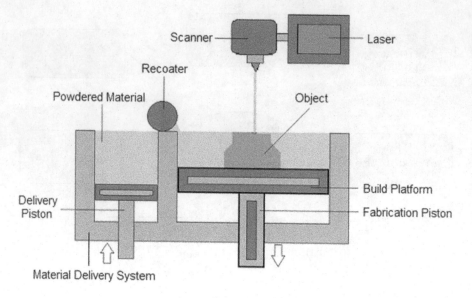

**FIGURE 4.2**   Illustration of the Working Principle of the SLM Process.

is executed in a vacuum environment within the system, in which a high-powered electron beam from an electron gun focuses across a thin layer of pre-laid powdered material and induces melting and re-solidification of the powder particles.

The establishment of a vacuum environment within the system is a time-consuming process, but once achieved, it is beneficial in various aspects. For example, a vacuum system can be readily heated and prevent the transfer of heat energy by convection that makes it easy to maintain a thermally stable environment and elevated temperature within the system. Consequently, the components fabricated by this process are relatively less distorted thermally. This process is associated with the capability to fabricate parts with a lower porosity and higher building rate. However, it offers poor surface finish and meagre dimensional accuracy. In the case of parts comprised of titanium-based alloy Ti-6Al-4V fabricated by EBM process shows poor surface finish (arithmetic roughness ~25 μm) that adversely affects the fatigue life of the part (Spitaels et al., 2020).

### 4.2.2 MATERIAL EXTRUSION

**Fused deposition modelling (FDM)** is a material extrusion-based AM technique. It is the second most important AM technique used commercially in AM domain across the globe. It is also recognised as the most economical AM technique. In this process, subsequent layers of the part are fabricated by extruding the filament material through a nozzle that follows the cross-sectional shape of the component. A resistive heater is installed with a nozzle that maintains the temperature of the filament near its melting point. An extruder is used to force out the filament material from the material spool. This process consists of a printable head that can move along the X and Y direction above the build platform, whereas the build platform moves along the Z direction.

Some of the important commercially used materials in FDM include PLA, ABS (acrylonitrile butadiene styrene) and a mixture of PC ABS (Bourell et al., 2017). FDM provides high dimensional accuracy and makes it easy to fabricate parts with complex geometries without using a die or mould. Figure 4.3 illustrates the basic process of FDM.

**Contour Crafting (CC)** is another material extrusion-based AM technique. It is mainly used in construction, in which material in form of paste is ejected against trowel that produces a uniform surface to stack subsequent layers. This process has precisely addressed the issue of high-speed automation construction (Lim et al., 2012). In this process, build material such as cement and mould material such as mortar mix is used.

### 4.2.3 Vat Photopolymerization

Vat photopolymerization is also known as stereolithography, in which photo-polymer material is kept in a movable vat. UV light selectively strikes over the photopolymer material in order to initiate a reaction which ultimately solidifies it, forming a thin layer. Figure 4.4 illustrates the main components of the

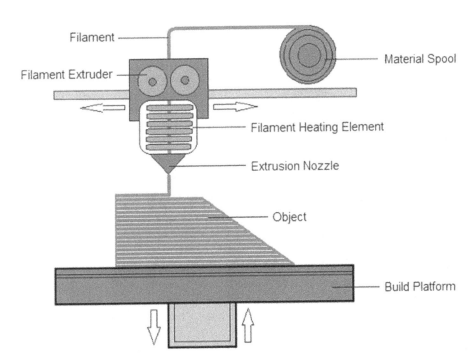

**FIGURE 4.3**   Illustration of the Working Principle of FDM.

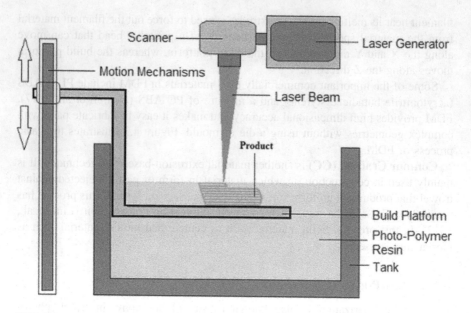

**FIGURE 4.4**  Illustration of The Working Principle of the Stereolithography Process.

stereolithography process of AM. Materials used in stereolithography include epoxies, resin and thermoset.

## 4.2.4  BINDER JETTING

In this process of AM, a liquid binding material is selectively spread over the layers of powdered material in order to fabricate a 3D component. This process offers a high building rate and rapid printing of parts with complex geometries. It can utilise a variety of powdered materials for manufacturing. Different types of materials such as polymers, metals and ceramics are used with BJ (Ziaee and Crane, 2019). In Figure 4.5, the working principle of a basic binder jetting process is depicted.

## 4.2.5  MATERIAL JETTING

Material jetting is similar to stereolithography. In this process of AM, photo-polymeric and build materials, in droplet form, are selectively deposited for the formation of a thin layer which is further cured by UV light. When the formation of one layer is completed, it is then lowered by a piston, and the formation of another layer is done over it. In this way, fabrication of components by material jetting process takes place (Li et al., 2019). In Figure 4.6, a neat sketch of material jetting process is shown. The commonly used materials in the material jetting process are thermoset photopolymers and resins.

This process provides smooth surface finish and high part accuracy and hence, it is utilised to fabricate high-detail scale models in the aerospace industry. For

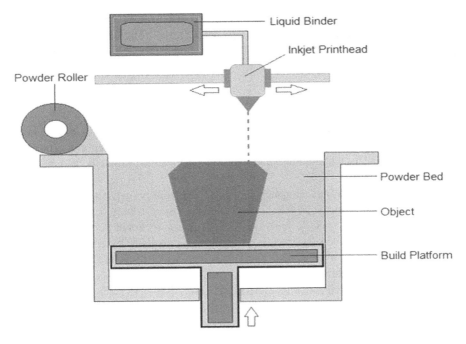

**FIGURE 4.5** Illustration of the Working Principle of the Binder Jetting Process.

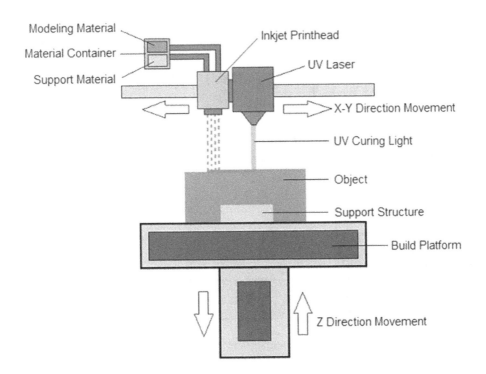

**FIGURE 4.6** Illustration of the Working Principle of the Material Jetting Process.

example, a fully transparent high-detail headlight prototype of an aerospace design is fabricated by material jetting using transparent resins. The material jetting process is also used for producing multi-material components of complex geometries for moulding and medical uses.

## 4.2.6 Directed Energy Deposition

Directed energy deposition (DED) is an AM technology that makes use of heat energy from a laser beam in order to selectively melt the metal when it is deposited to form a layer. This technique is suitable for the fabrication of metallic and metallic hybrid components.

**Laser engineered net shaping** is a directed energy deposition-based AM technique in which metal powders are melted by using a laser beam and subsequently sintered into specific locations to form cross-sectional geometries of the subsequent layers of the part. Each layer is allowed to solidify before the formation of a successive layer over it. This process is carried out in an inert gas shielding atmosphere to mitigate in-situ oxidation.

A variety of materials such as metals – Ti, 316 L stainless steel (Jackson et al., 2020), maraging steel, tool steel, aluminium alloy – AL5083, super alloy iconol 718 (Ning and Wang, 2017) and ceramic-Al2O3-ZrO2 (S. Yan et al., 2018) are used in this process. In Figure 4.7, the working principle of the laser-engineered net shaping process is illustrated.

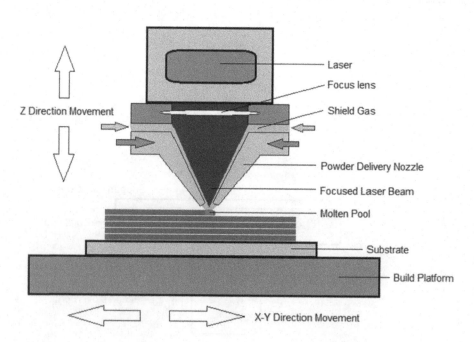

**FIGURE 4.7** Illustration of the Working Principle of the LENS Process.

This process provides high printing speed and produces less material waste. It produces highly dense components that have mechanical properties comparable to components produced by casting.

In various industries such as aerospace, automotive and architecture, DED is implemented in the manufacturing of components. Structural parts of fighter aircraft and satellites are produced using this process. In addition to the production of metallic components, DED is employed to repair damaged components, and it is also used to modify parts. The corrosive and resistive properties of items, such as turbine blades and injection tools, are enhanced by depositing a hard face layer over the surface of the component through DED.

### 4.2.7 Sheet Lamination

Laminated object manufacturing (LOM) is an AM technique that makes use of materials in the form of sheets. In this process, sheets are cut using a laser beam to be able to shape the layers of a 3D object. These sheets are joined together by applying a coating of thermal adhesive material between them, in combination with pressure and heat. In this process, the materials such as paper, metal sheets and composite metal sheets (Liu et al., 2018) are used to fabricate 3D objects (Figure 4.8).

Sheet lamination has various areas of applications. In this technique, metal sheets are utilised in the manufacturing of hybrid components, and a paper-based process such as LOM is used to produce multicolour objects. Table 4.1 presents the advantages and limitations of various AM based techniques.

## 4.3 MATERIALS IN AM

### 4.3.1 Plastic

Thermoplastic polymers are preferred in the material extrusion and powder bed fusion processes of additive manufacturing. For the material extrusion process,

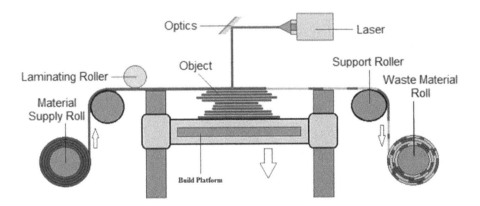

**FIGURE 4.8**   Illustration of the Working Principle of the Sheet Lamination Process.

**TABLE 4.1**

**Different Types of AM Techniques and their Materials, Advantages and Limitations**

| AM Technique | Material | Advantages | Limitations |
|---|---|---|---|
| powder bed fusion (PBF) | metals such as Ti, Al, ceramics, polymers, composites | • Easy removal of support structures is possible.<br>• Product with fine resolution is fabricated.<br>• Unused powder after fabrication can be reused. | • Building speed is slow.<br>• It produces products with a rough surface. |
| vat photopolymerization | ceramic materials such as alumina and photopolymers such as epoxies | • Parts with fine resolution can be fabricated.<br>• It gives high dimensional accuracy.<br>• Parts with intricate details can be fabricated. | • Relatively lesser materials are available.<br>• Parts are of poor mechanical properties. |
| sheet lamination | polymer-metal-ceramic hybrid | • It is a low-cost process.<br>• It does not require any support structure.<br>• There is no requirement for post-processing. | • Wastage of material is high.<br>• It is comparatively difficult to fabricate parts with cavities. |
| material jetting | polymer-metal-ceramic hybrid | • It gives high accuracy in droplet deposition.<br>• It produces low waste.<br>• It gives a good surface finish. | • This process is limited to photopolymers and thermoses. |
| binder jetting | polymer and ceramic-composite metal | • It fabricates objects with higher precision.<br>• It is the cheapest and fastest fabrication technique. | • This process is limited to low strength use.<br>• The mechanical properties of parts fabricated are mediocre.<br>• It gives poor surface finish. |
| material extrusion | thermoplastic such as ABS, PLA, PCL, PLLA, etc. and ceramic-composites-metals | • Parts with multi-material can be produced.<br>• It is an economical and simple fabrication AM technique. | • Products with poor surface quality are produced.<br>• Products have an anisotropic material structure. |
| direct energy deposition | metal and metal hybrid | • This process is suitable for repairs.<br>• It provides higher control of grain structure.<br>• It produces high-quality product. | • This process is limited to metals and metal composites.<br>• It provides a good balance between surface quality and print speed. |

amorphous thermoplastic polymers are suitable, and this includes ABS and PLA. These materials remain soft in a broad range of temperatures and maintain high viscosity which is suitable for material extrusion.

In the powder bed fusion process, UV rays are utilised for the melting and fusion of semi-crystalline powder. The most popular semi-crystalline material for powder bed fusion process is polyamide 12 (Nylon). The photopolymer materials are composed of monomer, oligomers, photo-initiators and a variety of additives, including dry and toughing agents. A mixture of acrylate monomer and UV photo-initiators was initially used in vat photopolymerization (Jacobs, 1996). Vinyl ether is another monomer that is used in the vat photopolymerization process. These materials exhibit 5% to 20% of volume shrinkage that causes residual stress in the parts (Crivello and Dietliker, 1998).

Epoxies are another class of thermoset materials that are used in the vat photopolymerization process. These polymers have a ring structure that opens in order to form new bonds during the chemical reaction. Ring-opening causes less volume shrinkage (Jacobs, 1996). During the polymerisation process, epoxy monomers act as plasticizers and consequently increase the mobility of molecules.

### 4.3.2 METALS

Metals are mainly used in powder bed fusion and the direct energy deposition process of AM. Different types of metals and metal alloys used in additive manufacturing include pure titanium, Ti6Al4V (Brandl et al., 2010), 316L stainless steel (Niendorf et al., 2013), 17-4 PH stainless steel, 18Ni300 maraging steel and nickel-based superalloys Inconel 718 (Amato et al., 2012).

Other alloys such as aluminium-based Al-Si-Mg 6001, tool steel-based H13 cermets, and superalloys N718 and Stellite are also used in AM (Uriondo et al., 2015).

### 4.3.3 CERAMIC AND COMPOSITE

Due to high melting point and low durability qualities, it is difficult to directly use ceramic materials in additive manufacturing (Clare et al., 2008). Although, Alumina and its alloys are used in the directed energy deposition process (Balla et al., 2008). Ceramics are generally used with binder materials. Composite materials utilised in the AM process mainly consist of polymer composites and metal composites.

## 4.4 APPLICATION

Additive manufacturing is the future of manufacturing industries. It has the potential to become the most economical way of manufacturing and has the potential to replace most of the traditional subtractive manufacturing processes. AM has already benefited many industries, including the automotive, aerospace, healthcare and electronics industries. Day-to-day developments in AM technologies are continuously increasing the number of manufacturing segments benefitting from these.

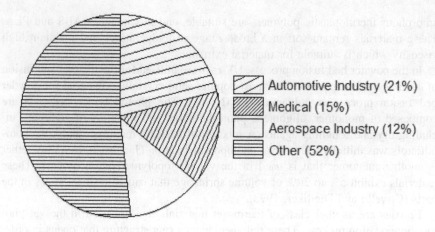

**FIGURE 4.9** Contribution of Various Industries in the Market Share of Additive Manufacturing.

According to a report published in 2017 by Wohler, the automotive, aerospace and medical industries contribute 48% of the total market share of additive manufacturing. In Figure 4.9, the individual contribution of these three industries in the total market share of AM is represented by a pie chart (Wohlers, 2017). Although, the contribution of the construction industry in the total market share of AM is not significant, there are boundless possibilities for AM in this field. Based on this, the applications of AM in the automotive, aerospace, medical and construction industries are briefly summarised.

### 4.4.1 Aerospace Industry

AM is one of the important fabrication technologies adopted in the aerospace industry. AM generates about 18.2% of its total revenue from this industry (Najmon et al., 2019). In this field, AM technologies are capable of fulfilling the demands of light aircraft parts with a high mechanical strength used to enhance fuel efficiency and reduce emission. (Braga et al., 2014).

The application of additive manufacturing technologies in the aerospace industry can be generally classified into the following categories: direct digital manufacturing (DDM), rapid tooling, rapid prototyping and repair.

DDM refers to the fabrication of parts that are directly used in aircraft. These parts are important when considering the safety and reliability of aircraft and hence, must be fabricated with critical care. Rapid tooling ensures the production of mechanisms and parts that are used for the fabrication of the aircraft component. It is described as any mould-making process that creates tools in a short interval of time by employing minimum labour. Rapid prototyping refers to the fabrication of parts that are non-functional and usually fabricated using nonmetals. Repair refers to the overhaul and maintenance of metal parts and connections usually, by using DED and certain cold spray processes. Repairing by additive technology induces less environmental effects compared to other conventional technologies (Walachowicz et al., 2017).

In the aerospace industry, both metallic and nonmetallic components are fabricated by additive manufacturing. The door handle of Boeing 777-300ER (Hipolite, 2015) and camera case prototyping (Moore, 2018) are some examples of nonmetallic components, whereas the combustion chamber of a helicopter engine (Hauser, 2014) and lap joint reinforcement (Jones et.al., 2014) are the examples of metallic components constructed using additive manufacturing. Table 4.2 shows a breakdown of the application of AM technologies in the aerospace industry.

### 4.4.2 AUTOMOTIVE INDUSTRY

In the automotive industry, there are enormous opportunities for additive manufacturing. Many vehicle parts are already being manufactured by this ideal and favoured process and it is predicted to expand further.

The expansion of AM in the automotive industry is the result of comparatively less manufacturing cost, quick production and reduction of the lengthy design process of the parts that may be produced by AM. It also ensures the faster availability of the parts to customers as well as rapid customisations that need to be implemented in the components.

There are many techniques of AM that are utilised for the production of a number of different types of components in the automotive industry. Some of these are listed in Table 4.3.

### 4.4.3 CONSTRUCTION INDUSTRY

AM technologies can be advantageous in large-scale construction because it has the ability to realise a complex architectural structure on the ground. AM offers several benefits, including the removal of the replication of components as well as the unique design of each component. AM processes that are being used commercially in construction and architecture include concrete printing (CP), contour crafting (CC), D-shape (monolite) (Lim et al., 2012). All of the three processes are successfully used in the construction of civil structures and are favourable for

---

**TABLE 4.2**

**Various Parts Manufactured by AM-based Techniques in the Aerospace Industry**

| Material | AM Technique | Example |
|----------|--------------|---------|
| metal | binder jetting technique | lap joint reinforcement (repairing) |
| | direct energy deposition based technique | fabrication of combustion chamber for helicopter engine (DDM) |
| nonmetal | material extrusion-based FDM technique | door handle of Boeing 777-300ER (DDM) |
| | material extrusion-based FDM technique | camera case prototyping (rapid prototyping) |

**TABLE 4.3**

**Applications of AM in the Automotive Industry**

| Area | Parts | Ink Material | Process |
|------|-------|--------------|---------|
| fluid handling | pump and valve | aluminium alloy | It is fabricated by powder bed fusion techniques, such as EBM and SLM. |
| exterior | windbreakers and bumpers | polymers | It is fabricated by the powdered bed fusion-based SLS process. |
| exhaust and emission | cooling vents | aluminium alloy | It is fabricated by the powdered bed fusion-based SLM process. |
| manufacturing process | prototype, casting and customised tooling | hot work steel with polymers | For this material, extrusion-based FDM and powdered bed fusion-based SLS and SLM processes are used. |

construction and architectural work. These processes additionally build the components; however, each process is developed for their specific application and material requirements.

**Contour Crafting** is one of the first AM technologies adopted in construction. It is an extrusion-based AM process in which the extrusion of material paste, such as concrete, is achieved by a print head (Zhang and Khoshnevis, 2010). A trowel is used for smoothening the surfaces created by CC. This process requires support structures for the construction of overhangs and is appropriately used in small structures and housing construction work.

**D-Shape** is a powder deposition-based AM process in which a layer of desired depth and compactness of building material, such as sand, is created. Next, the binding material is selectively deposited in order to solidify the layer of building material by using a nozzle mounted over a gantry frame.

**Concrete printing** is a type of important AM technology used in construction. In this process, the print head extrudes the mixture in a predefined path in order to build the structure. The print head consists of a nozzle that varies from 4 mm to 9 mm in diameter (Lim et al., 2009). In this process, the trowel is not used for smoothening surfaces. Rather, it constructs structures with rough surfaces. In this process, the extrusion of cement mortar is carried out without compromising the three-dimensional freedom of the process, and it has a high resolution of deposition, which gives greater control of internal and external geometries. Concrete printing print resolution varies from 4 to 6 mm in terms of layer depth (Dooil and Behrokh, 2004).

In all three processes, the hardening of material is carried out by the curing process which is less controllable. Concrete printing and contour crafting are wet processes, whereas the D-shape process is a dry process (Craveiro et al., 2019). A comparison of these three processes is shown in Table 4.4.

**TABLE 4.4**

**Comparison of Contour Crafting, Concrete Printing and D-Shape Techniques used in the Construction Industry**

| Process | Material | Advantages | Limitations |
|---|---|---|---|
| D-shape | building material: sand, stone powder binding material: chlorine-based liquids | It can create structures with complex geometries. | It is comparatively slower than CC and CP. |
| concrete crafting (CC) | ceramic paste, concrete | It generates smooth surfaces, and it can give higher building rates. | It is limited to vertical extrusion, and it is suitable for small structure construction. |
| concrete printing (CP) | mortar, gypsum | It can produce complex structures, and it also gives a higher building rate. | It creates structures with rough surfaces. |

### 4.4.4 HEALTHCARE INDUSTRY

AM technology plays an important role in the healthcare sector, particularly in the field of tissue engineering for fabricating scaffolds and matrices and in the pharmaceutical industry. Some of the major areas of the application of AM include the fabrication of anatomy models, mechanisms such as surgical instruments and pharmaceutical products (Trenfield et al., 2018). Different types of AM techniques are used in tissue engineering, depending on the requirement of materials. Furthermore, AM can be utilised for the fabrication of human organs and tissues in order to replace or repair damaged ones. A scaffold is a porous 3D structure that replaces damaged tissue and promotes the biological activities to prepare for the functioning of the cell. It consists of overly complex geometry making it arduous to produce, but AM technology has the potential to fabricate it.

Materials used for the fabrication of scaffolds must be biocompatible, biodegradable and non-toxic. Biocompatibility of material ensures that it does not initiate an inflammatory reaction nor exhibit cytotoxicity when comes in contact with human tissue. Biodegradability of material provides growth of natural tissues with the degradation of scaffold material. Large varieties of materials that are suitable for scaffold fabrication are available. This includes metal alloys of Ti and Mg, biologically active ceramics, composites made up of ceramics and polymers and certain natural and synthetic polymers such as hydrogels, thermoplastic, protein and elastomeric of thermoplastic (Chen et al., 2012).

In pharmaceutical industries, additive manufacturing techniques are used to assemble tablets in different shapes including the sphere, pyramid and torus. FDM is one of the additive manufacturing technologies which are appropriately used for this purpose. Additive manufacturing makes it possible to manufacture tablets consisting of distinct regions (on one tablet) containing different types of drugs. In this manner, additive manufacturing is adopted for the fabrication of polypills in pharmaceutical industries.

## 4.5 FUTURE TRENDS IN AM

Many companies have realised the need for transformation in traditional manufacturing in order to improve the quality and quantity of products. With the developments in technology, the call for new and innovative products is also emphasised. AM has the potential to play a vital role in achieving these requirements. Some of them are further discussed below.

AM offers large geometric freedom to the fabrication of the products. Therefore, it has the potential to be utilised for developing a multifunctional structure that is capable of executing multiple tasks in a single time frame. There are several examples of AM, such as the creation of products that constitute both soft and rigid material and the manufacturing of structures with embedded electronics, where it has performed exceptionally well.

Four-dimensional printing is another elegant feature of future manufacturing where the structure will be capable of changing its shape and size in accordance with external environmental parameters such as temperature, moisture content, radioactivity, and many more (Kwok et al., 2015). Polymer products are sensitive to environmental parameters, like temperature, and can be utilised in adding and removing assemblies according to external environmental conditions (Teoh et al., 2017).

The failure of high-value components can cause adverse effects because repairs are always an expensive and time-consuming task. AM is flexible enough to develop an automated repairing system so that, in a short interval of time and with fewer expenses, high-value components can be repaired and failure can finally be prevented.

The optimisation of product design in terms of weight, strength, stiffness and many other parameters can be achieved using multiple materials in the manufacturing of products, and AM is capable of using multiple materials to be able to fabricate a product.

With the developments in AM technology, it is feasible to fabricate larger components, such as aircraft wings, in the near future (Liu et al., 2017). AM is also capable of repairing such massive components in a significantly short lead time and with small expenses. With the development of new AM materials that possess competent mechanical and thermal properties, the use of AM in various industries will accelerate and improve the future of AM technology in the world of manufacturing.

## 4.6 CONCLUSION

This chapter presented a comprehensive review on various additive manufacturing techniques which are being used commercially in various manufacturing areas. AM technologies are responsible for the innovation and improvement in the traditional methods of manufacturing in combination with the reduction in the wastage of materials, energy and optimisation of product design. We have discussed the challenges in the present scenario of the manufacturing world along with recent developments in AM that make it feasible for application. An analysis of AM reveals that developments in these technologies have sped up in recent years, and it is the consequence of their increasing area of applications, synthesis of a wide range of materials that are in line with AM technologies and the research work of the experts for the improvement of such a unique manufacturing tool. Furthermore, we have discussed the applications of AM technologies in four major areas of interest where it is more prominently welcomed as compared to others. This includes the automobile, construction, healthcare and aerospace industries. Finally, we have analysed how recent developments are shaping the future of AM technology in the world of manufacturing.

## REFERENCES

Amato, K. N., S. M. Gaytan, L. E. Murr, E. Martinez, P. W. Shindo, J. Hernandez, S. Collins, F. Medina. 2012. Microstructures and mechanical behaviour of Inconel-718 fabricated by selective laser melting. *Acta Materialia* 60:2229–2239.

Balla, V. K., S. Bose, A. Bandyopadhyay. 2008. Processing of bulk alumina using laser engineered net shaping. *International Journal of Applied Ceramic Technology* 5:234–242.

Bourell, D. Jean Pierre Kruth, Ming Leu, Gideon Levy, David Rosene, Allison M. Beese, Adam Clare, 2017. Materials for additive manufacturing. *CIRP Annals- Manufacturing Technology* 66:659–681.

Braga, D. F. O., S. M. O. Tavares, Lucas F. M. da Silva, P. M. G. P. Moreira, Paulo M. S. T. de Castro. 2014. Advanced design for light weight structure: Review and prospect. *Progress in Aerospace Sciences* 69:29–39.

Brandl, E., B. Baufeld, C. Leyens, R. Gault. 2010. Additive manufactured Ti-6Al-4v using welding wire: Comparison of laser and arc beam deposition and evaluation with respect to aerospace material specifications. *Physics Procedia* 5:595–606.

Chen, Q., C. Zhu, G. A. Thouas. 2012. Progress and challenges in biomaterial used for bone tissue engineering: bioactive glasses and elastomeric composites. *Progress in Biomaterials* 1. http://dx.doi.org/10.1186/2194-0517-1-2.

Clare, A. T., P. R. Chalker, S. Davies, C. J. Sutcliffe, S. Tsopanos. 2008. Selective laser sintering of barium titanate-polymer composite films. *Journal of Materials Science* 43:3197–3202.

Craveiro, Flavio, Jose Pinto Duarte, Helena Bartolo, Paulo Jorge Bartolo. 2019. Additive manufacturing as an enabling technology for digital construction: A perspective on construction 4.0. *Automation in Construction* 103:251–267.

Crivello, J. V., K. Dietliker. 1998. Photoinitiators for free radicals, cationic & anionic photopolymerization. In Bradley G., (Ed.) *Chemistry and Technology UV & EB Formulation for Coatings Inks & Paints.*. Chichester, United Kingdom: John Wiley & Sons.

Dooil, Hwang, Khoshnevis Behrokh. 2004. Concrete wall fabrication by contour crafting. Paper Presented at the 21st International Symposium on Automation and Robotics in Construction, South Korea.

Gibson, Ian, David W. Rosen, Brent Stucker. 2010. *Additive Manufacturing Technologies. Rapid Prototyping to Direct Digital Manufacturing.* MA: Springer US.

Halloran, J. W., V. Temeckova, S. Gentry et al. 2011. Photopolymerization of powder suspension for shaping ceramics. *Journal of European Ceramic Society* 31:2613–2619.

Hauser, C. 2014. *Case Study: Laser Powder Metal Deposition Manufacturing of Complex Real Parts*. Cambridge, United Kingdom: The Welding Institute.

Hipolite, W. 2015. China Eastern airlines successfully 3D prints airplane parts for Boeing 777-300ER Aircraft. *3DPRINT.COM*. https://3dprint.com/57751/china-eastern-airlines-3 d-print/.

Jackson, M. A., Justin D. Morrow, Dan J. Thoma, Frank E. Pfefferkorn. 2020. A comparison of 316 stainless steel parts manufactured by directed energy deposition using gas-atomized and mechanically-generated feedstock, *CIRP Annals – Manufacturing Technology* 69:165–168. https://doi.org/10.1016/j.cirp.2020.04.042.

Jacobs, P. F. 1996. *Stereolithography and Other RP&M Technologies: From Rapid Prototyping to Rapid Tooling*. Dearborn, MI: Society of Manufacturing Engineers.

Jones, R., L. Molent, S. Barter, N. Matthews, D. Tamboli. 2014. Supersonic particle deposition as a means for enhancing the structural integrity of aircraft structures. *International Journal of Fatigue* 68: 260–268.

Kwok, Tsz-Ho, Charlie C. L. Wang, Dongping Deng, Yunbo Zhang, Yong Chen. 2015. Four-dimensional printing for freeform surfaces: Design optimization of origami and kirigami Structures. *Journal of Mechanical Design* 137:111712–111722.

Li, V. C. F., X. Kuang, C. M. Hamel, D. Roach, Y. Deng, H. J. Qi. 2019. Cellulose nano-crystal support material for 3D printing complexly shaped structures via multi-material-multi-printing. *Additive Manufacturing* 28:14–22.

Liu, H., Wenhao Zhang, Jenn-Terng Gau, Zongbao Shen, Guoce Zhang, Youjuan Ma, Xiao Wang. 2018. Microscale laser flexible dynamic forming of Cu/Ni laminated composite metal sheets. *Journal of Manufacturing Processes* 35:51–60.

Liu, R., Z. Wang, T. Sparks, F. Liou, J. Newkirk. 2017. Aerospace applications of laser additive manufacturing, *Laser Additive Manufacturing, Woodhead Publishing Series in Electronic and Optical material* 351–371.

Lim, S., T. Le, J. Webster, R. Buswell, S. Austin, T. Thorpe. 2009. Fabricating construction components using layer manufacturing technology. Paper Presented at Global Innovation in Construction Conference, Loughborough University, UK.

Lim, S., R. A. Buswell, T. T. Le, S. A. Austin, A. G. F. Gibb, T. Thorpe. 2012. Developments in construction-scale additive manufacturing processes. *Automation in Construction* 21:262–268.

Matta, A. K., Shyam Prasad Kodali, Jayant Ivvala, P. Jamaleswara Kumar. 2018. Metal prototyping the future of automobile industry: A review. *Material Today: Proceedings* 5:17597–17601.

Moore, S. 2018. 3-D printing deployed in aerospace sector for tooling, prototyping. *Plastics Today*. https://www.plasticstoday.com/3d-printing/3d-printingdeployed-aerospace-sector-tooling prototyping/32485232158131.

Najmon, Joel C., Sajjad Raeisi, Andres Tover. 2019. Review of additive manufacturing technologies and application in aerospace industry. *Additive Manufacturing for Aerospace Industry* 1:7–31.

Niendorf, T., S. Leuders, A. Riemer, H. Richard, T. Tröster, D. Schwarze. 2013. Highly anisotropic steel processed by selective laser melting. *Metallurgical and Materials Transactions* 44:794–796.

Ning, Fuda, Xinlin Wang. 2017. Ultrasonic vibration-assisted laser engineered net shaping of Iconol 718 parts: A feasibility study. *Procedia Manufacturing* 10:771–778.

Phillips, Tim, Scott Fish, Joseph Beaman. 2018. Development of an automated laser control system for improving temperature uniformity and controlling component strength in selective laser sintering. *Additive Manufacturing* 24:316–322.

Puma-Araujo, S., D. Olvera-Trejo, A. Elías-Zuñiga, O. Martinez-Romero, C. A. Rodríguez. 2016. Design and characterization of a magnetorheological damper for vibration mitigation during milling of thin components. *MRS Online Proceedings Library* 65–70. https://doi.org/10.1557/opl.2016.19.

Stichel, Thomas, Thomas Frick, Tobias Laumer, Felix Tenner, Tino Hausotte, Marion Merklein, Michael Schmidt. 2018. A round robin study for selective laser sintering of polymers: Backtracing of the pore morphology to the process parameters. *Journal of Material Processing Technology* 252:537–545.

Spitaels, Laurent, François Ducobu, Anthonin Demarbaix, Edouard Rivière-Lorphèvre, Pierre Dehombreux. 2020. Influence of conventional machining on chemical finishing of Ti6Al4V electron beam melting parts. *Procedia Manufacturing* 47:1036–1042.

Sudarev, A., V. Konakov, Y. Chivel. 2018. Selective laser sintering of ceramic turbo machine components. *Procedia CIRP* 74:264–267.

Teoh, Joanne Ee Mei, Yue Zhao, Jia An, Chee Kai Chua, Yong Liu. 2017. Multi-stage responsive 4D printed smart structure through varying geometric thickness of shape memory polymer. *Smart Materials and Structures* 26:125001.

Trenfield, S. J., C. M. Madla, A. W. Basit, S. Gaisford. 2018. The shape of things to come: emerging applications of 3D printing in healthcare. *3D Printing of Pharmaceuticals* 31:1–19.

Uriondo, A., M. Esperon-Miguez, S. Perinpanayagam. 2015. The present and future of additive manufacturing in the aerospace sector: A review of important aspects. *Proceedings of the Institution of Mechanical Engineers, Part G: Journal of Aerospace Engineering* 229. https://doi.org/10.1177/0954410014568797.

Walachowicz, F., Ingo Bernsdorf, Ulrike Papenfuss, Christine Zeller, Andreas Graichen, Vladimir Navrotsky, Noorie Rajvanshi, Christoph Kiener. 2017. Comparative energy, science, recycling lifecycle analysis in industrial repair process of gas turbine burners using conventional machining and additive manufacturing. *Journal of Industrial Ecology* 21:203–215.

Wohlers, T. 2017. *3D Printing and Additive Manufacturing State of the Industry.* Annual Worldwide Progress Report. Wohler's Associates: Fort Collins, CO.

Yan, Q., B. Chen, N. Kang, X. Lin, S. Lv, K. Kondoh, S. Li, J. S. Li. Comparison study on microstructure and mechanical properties of Ti-6Al-4V alloys fabricated by powder-based selective-laser-melting and sintering methods, *Materials Characterization* 164:110358. https://doi.org/10.1016/j.matchar.2020.110358.

Yan, Shuai, Dongjiang Wu, Fangyong Niu, Yunfei Huang, Ni Liu, Guangyi Ma. 2018. Effect of ultrasonic on forming quality of nano sized Al2O3-ZrO2 eutectic ceramic via laser engineered net shaping (LENS). *Ceramics International* 44:1120–1126.

Zhang, J., B. Khoshnevis. 2010. Contour crafting process plan optimization part I: Single nozzle case. *Journal of Industrial and Systems Engineering* 4:33–46.

Ziaee, Mohsen, Nathan B. Crane. 2019. Binder jetting: A review of process, materials, and methods. *Additive Manufacturing* 28:781–801.

Zindani, Divya, Kaushik Kumar. 2019. An insight into additive manufacturing of fibre reinforced polymer composite. *International Journal of Lightweight Materials and Manufacture* 2:267–278.

# 5 Advancements in Post-Processing of Metal Additive Manufactured Components

*A. N. Jinoop, C. P. Paul, and K. S. Bindra*

## CONTENTS

## 5.1 INTRODUCTION

Additive manufacturing (AM) has grown significantly in the last three decades in terms of the number of machines sold and parts produced. Also, a steady increase in the number of publications and patents reiterates the advancements in AM. Figure 5.1 presents the increase in the number of publications in the ScienceDirect database from 2000 to 2018 (du Plessis, Broeckhoven, et al. 2019). The contribution of AM towards functional components increased from 28% in 2012 to 34% in 2016 (Gisario et al. 2019). These expansions are mainly due to the advances in metal AM (MAM). MAM transformed the technology to a stage where it can build seamless components with all of the necessary functions that are required for a particular application.

The classification of MAM can be done based on the type of raw material, raw material feeding and energy source, as presented in Figure 5.2. Powder and wire-based MAM are the two types depending on the type of raw material used. Based on

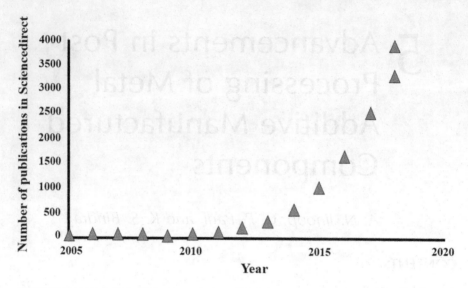

**FIGURE 5.1**  Yearly Distribution of AM Publications from Sciencedirect. (Reprinted from Additive Manufacturing, 27, Plessis, Anton du, Chris Broeckhoven, Ina Yadroitsava, Igor Yadroitsev, Clive H Hands, Ravi Kunju and Dhruv Bhate, Beautiful and Functional: A Review of Biomimetic Design in Additive Manufacturing, 408–427, 2019, with kind permission from Elsevier).

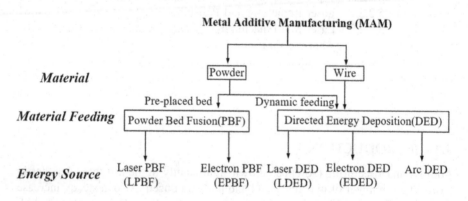

**FIGURE 5.2**  Classification of MAM Processes.

the material feeding technique, powder-based MAM can be classified into pre-placed powder bed technique and dynamic feeding technique. A type of MAM technique that uses pre-placed powder bed technique is called powder bed fusion (PBF), as shown in Figure 5.3 and those techniques that deploy dynamic material feeding is called directed energy deposition (DED) (Paul, Jinoop, and Bindra 2018), as shown in Figure 5.4. Wire-based MAM solely follows the dynamic feed technique. Based on the type of energy source, MAM can be classified into laser beam-based MAM, electron beam-based MAM and arc-based MAM. Laser beam-based MAM is generally used in PBF and DED configurations as LPBF and LDED, respectively. Electron beam MAM is

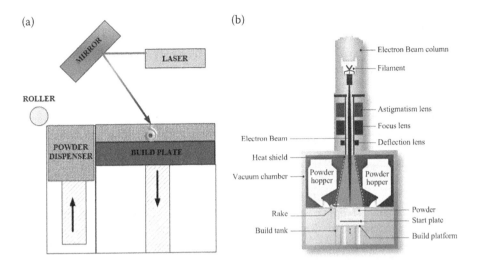

**FIGURE 5.3** General Schematic of (a) LPBF (b) EPBF process (Reprinted from Additive Manufacturing, 19, Data from Galati, Manuela and Luca Iuliano A Literature Review of Powder-Based Electron Beam Melting Focusing on Numerical Simulations, 1–20, 2018, with kind permission from Elsevier).

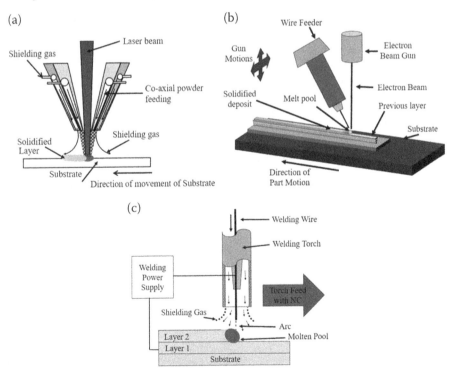

**FIGURE 5.4** General Schematic of (a) Powder-Based LDED (b) Wire-Based DED (d) Arc-based DED.

also available in PBF and DED configurations as EPBF and EDED, while arc-based MAM is available only in the DED configuration. LDED can use powder or wire as the feed material, while arc and electron-based DED uses only wire as the raw material (Gisario et al. 2019). Table 5.1 presents a comparison of the advantages, disadvantages and commercial names of PBF and DED processes.

Even though MAM techniques have several advantages, there are many defects and limitations for MAM-built components. The major flaws of MAM-built components are porosity, surface roughness and waviness, cracking, delamination and distortions.

Porosity is generated either due to improper process parameter selection or the powder generation step. Surface roughness is higher for MAM-built components as compared to conventionally manufactured parts due to the stair-stepping effect / Christmas tree effect and the presence of partially melted particles on the built surface. The stair-stepping effect is due to the layer-by-layer fabrication, which results in a deviation of the built surface from the ideal surface. The presence of partially melted particles will happen due to the incorrect selection of different process parameters. Waviness is mainly due to the laser scan pattern used for deposition. MAM-built components are subjected to cracking and these can be mainly of two types:

- Solidification cracking: Solidification cracking happens when the energy used to melt the material is too high, and this results in a large amount of stress generation between the solidified region and regions that are yet to be solidified. This type of cracking is a primary function of the pattern of material solidification and is classically instigated by high strain on the melt pool, insufficient melt pool flow or flow hindrance by the solidified grains.
- Grain boundary cracking: These cracks generally nucleate along the material grain boundaries and are material dependent. These depend on the generation/

## TABLE 5.1
## Comparison between PBF and DED Processes

| Process | Advantages | Disadvantages | Commercial Name |
|---|---|---|---|
| PBF (refer Figure 5.3) | • complex geometry<br>• thin features<br>• small footprint<br>• lightweight designs | • relative slowness<br>• size limitations<br>• distortions<br>• limited multi-material options | • selective laser melting<br>• selective laser sintering<br>• direct metal laser sintering<br>• electron beam melting |
| DED (refer Figure 5.4) | • high deposition rate<br>• possibility of building large components<br>• more multi-material options | • limited complexity<br>• difficulty in achieving micron level features<br>• high energy usage | • laser engineered net shaping<br>• laser metal deposition<br>• direct metal deposition<br>• electron beam additive manufacturing<br>• wire arc additive manufacturing |

dissolution of precipitate phases and the grain boundary morphology (Sames et al. 2016).

Cracking is generally microscopic and is eliminated by tailoring processing conditions or post-processing. Delamination is a macroscopic defect by which a layer becomes fully or partly separated from the remaining layers due to the higher residual stresses or incomplete bonding with the previously-built layer. It cannot be eliminated by using post-processing techniques. Residual stresses generated during MAM can also result in distortion in the built components and deviations from the required dimensions (Sames et al. 2016).

The abovementioned defects limit the use of MAM-built components for various applications in the as-built condition. To make MAM-built components suitable for various applications, it is necessary to improve the surface topography, mechanical properties, surface properties and density and achieve uniform microstructure. This is done in the final stage of MAM, i.e. post-processing. Post-processing is the stage at which the as-built samples are subjected to various treatments, as per the final user requirements. It can be classified mainly into aesthetic, property and surface quality improvement techniques (Gibson et al., 2010)

The most commonly used post-processing techniques for MAM-built samples are heat treatment, computer numerical control (CNC)-based machining and hot-isostatic pressing (HIP). Heat treatment is typically used to achieve uniform microstructure, stress-relieving, and generation of strengthening phases. CNC machining is the most common finishing technique that is used for MAM-built components to improve the surface quality and dimensional accuracy. HIP is the post-processing technique that is generally applied to PBF built components using high pressure and temperature to enhance the density and mechanical properties (Sames et al. 2016; Paul, Jinoop, and Bindra 2018).

This chapter reviews the various post-processing techniques that are used for post-processing MAM-built components with a focus on various advanced techniques. These post-processing strategies are classified into two major categories: residual stress modification techniques and surface property improvement techniques. The residual stress modification techniques discussed in this chapter are laser peening and ultrasonic peening. In addition to tailoring the surface residual stresses, these also improve the surface properties. The surface property modification techniques are laser remelting and polishing, laser micromachining and abrasive finishing. Process physics will be discussed along with their applications for MAM-built components and the latest research developments.

## 5.2 POST-PROCESSING TECHNIQUES

### 5.2.1 RESIDUAL STRESS MODIFICATION TECHNIQUES

#### 5.2.1.1 Laser Shock Peening

Laser shock peening (LSP) is an advanced version of the shot peening process, in which a laser, instead of hardened balls, is used to induce compressive residual stress on the component surface. LSP uses a laser with a power density greater than

1 GW cm$^{-2}$ and a short pulse duration (in nanosecond range). Even though the laser is used in LSP, the process is purely mechanical and not thermal (Gujba and Medraj 2014; Yadav et al. 2017; Wang et al. 2019). Figure 5.5(a) presents the typical schematic of an LSP process, which essentially consists of a transparent layer (running water or glass), an opaque layer coated on the sample surface (black tape, black paint or aluminium foil) and a pulsed laser. Generally, a pulsed Nd:YAG laser is used for LSP. The opaque layer is generally used to prevent thermal effects on the sample surface. The laser beam passes through the transparent layer and strikes on the opaque layer. Upon striking the opaque layer (denoted by black tape or paint), it vaporises and gets converted into dense plasma by continuously absorbing laser energy. The plasma expands quickly after absorbing the laser energy, but the transparent water layer prevents its expansion. This generates a high-pressure shock wave, which acts on the metal surface. If the peak pressure of the generated shock wave exceeds the dynamic yield strength of the material, then the material undergoes plastic deformation. This generates a high magnitude of compressive residual stress on the component surface, which can improve the fatigue life of a component as it can aid in crack closure. The crack closure due to the compressive stress will act as a driving force used to prevent fatigue crack propagation. The major advantage of LSP over shot peening is that the magnitude of compressive stress in LSP is higher at the surface and depth than conventional shot peening. LSP also induces compressive stresses to a greater depth than shot peening. In addition, LSP produces lower surface roughness than shot peened samples due to the extreme and multi-axial loadings in shot peening as compared to the uniaxial loading in LSP. LSP is used to improve fatigue life, stress corrosion cracking/corrosion resistance, surface hardness, oxidation resistance and wear resistance of metallic components. LSP can be applied to complex geometries as it uses a laser beam, which can be directed to any point using laser beam delivery technology (Gujba and Medraj 2014; Danduk et al. 2016; Yadav et al. 2017).

LSP is applied to different MAM built alloys, like LPBF built Inconel 718 (Jinoop et al. 2019), SS 316L (Kalentics, Boillat, Peyre, Ćirić-Kostić, et al. 2017;

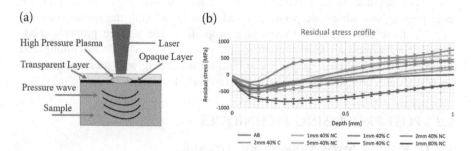

**FIGURE 5.5** LSP process (a) Schematic (b) the Effect of Process parameters on Residual Stress (Reprinted from Additive Manufacturing, 16, Kalentics, Nikola, Eric Boillat, Patrice Peyre, Snežana Ćirić-Kostić, Nebojša Bogojević and Roland E Logé, Tailoring Residual Stress Profile of Selective Laser Melted Parts by Laser Shock Peening, 90–97, 2017, with kind permission from Elsevier).

Kalentics, Boillat, Peyre, Gorny, et al. 2017; Hackel et al. 2018), 15-5 stainless steel (Kalentics, Boillat, Peyre, Ćirić-Kostić, et al. 2017), Ti-6Al-4V (Lu et al. 2020) and AlSi10Mg (du Plessis, Glaser, et al. 2019), LDED built Ti-6Al-4V (Guo et al. 2018) and arc-based deposition built 2319 aluminium alloy (Sun et al. 2018). Figure 5.5(b) presents the effect of transparent layer thickness and overlap percentage with or without opaque coating on the residual stress. The compressive stresses increases with an increase in the overlap percentage, presence of opaque coating, and reduction in the transparent layer thickness. As observed in shot peening, LSP also refines the grains near the surface and boosts the surface hardness and tribological behaviour of MAM-built components (Kalentics, Boillat, Peyre, Ćirić-Kostić, et al. 2017). For instance, a reduction of grain size from 33.6 to 24.3 μm is observed in LDED built Ti-6Al-4V (Guo et al. 2018).

Generally, LSP is only performed on the surface of MAM-built components. One of the recent advances is the 3D LSP, in which LSP is not only performed on the top surface, but also on the intermediate layers. Kalentics, Boillat, Peyre, Gorny, et al. 2017 reported the 3D LSP process on SS 316L, in which LSP is performed after building a few layers during LPBF. The major motivation behind 3D LSP is to achieve a larger depth of compressive stress for a given magnitude as presented in Figure 5.6(a), which compares the residual stress distribution in as-built, shot-peened, LSP processed and 3D LSP processed samples. The higher depth of compressive stresses in 3D LSP will aid in increasing the fatigue limit of LPBF built components. Figure 5.6(b) illustrates the depth variation of compressive residual stress for the as-built sample, the general LSP processed sample and the 3D LSP sample. 1 l, 10 l, and 30 l represented in Figure 5.6(b) show the number of layers after which the samples are subjected to LSP at a constant value of 1 mm transparent layer thickness and 40% overlap. The magnitude and depth

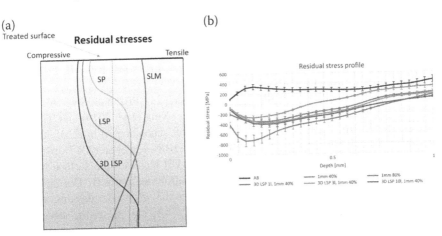

**FIGURE 5.6**  3D LSP (a) Typical Residual Stress Distribution (b) Residual Stress Profile at Different Conditions (Reprinted from *Materials & Design* 130, Kalentics, Nikola, Eric Boillat, Patrice Peyre, Snežana Ćirić-Kostić, Nebojša Bogojević and Roland E Logé, 3D Laser Shock Peening – A New Method for the 3D Control of Residual Stresses in Selective Laser Melting, 350–356, 2017, with kind permission from Elsevier).

of compressive stresses lessen with a reduction in the number of layers after which LSP is performed. Recently, Lu et al. (2020) investigated a different approach of 3D LSP by deploying LSP without coating for the peening of intermediate layers and LSP with coating on the top layer during LPBF of Ti-6Al-4V. The primary reasons for deploying LSP without opaque coating on intermediate layers are to induce compressive residual stress on the subsurface of a layer and to improve surface topography by removing some material during LSP without coating. Final LSP with opaque coating on the top surface induces compressive residual stress on the top layers. The method improves the tensile behaviour and surface hardness of the built structures with higher values of compressive stress on the top surface for the peened sample as compared to that of as-built samples without peening.

### 5.2.1.2  Ultrasonic Peening

Ultrasonic peening (UP) is another peening method used in the aerospace, automotive, railway, marine sectors, etc. in order to improve the surface hardness, wear resistance, corrosion resistance and fatigue life of components by inducing compressive residual stress on the surface. The major difference between UP and shot peening is the technique that is used to push the shots. Shot peening uses pneumatic forces to carry hardened balls to the sample surface. While UP uses the oscillating hard body to induce plastic deformation. Oscillating frequency in the ultrasonic range is used to impart plastic deformation to the sample surface. The typical frequency value used for UP is close to 27,000 Hz. Generally, the UP system consists of an ultrasonic signal generator, a transducer to convert the generated ultrasonic signals to mechanical motion and a metallic rod or horn that pushes the shots onto the metal surface (Yin et al. 2017). During UP, low amplitude high-frequency ultrasonic oscillations are applied to the surface through cylindrical pins or rods. This brings about plastic deformation on the metal surface, which leads to grain size refinement and compressive residual stress on the surface. The compressive residual stress on the sample surface closes the micro-cracks and prevents the propagation of cracks by eliminating tensile residual stress from the component surface (Malaki and Ding 2015).

UP is applied to different MAM-built alloys, such as LPBF built SS 316L (Joshua and Ajit 2017), NiTi (Ma et al. 2017), AlSi10Mg (Xing et al. 2019) and Ti-6Al-4V (Zhang et al. 2016), LDED built GH4169 (a nickel superalloy) (Liu et al. 2018) and arc-based deposition built Ti-6Al-4V (Yang et al. 2018). Ma et al. (2017) investigated UP on LPBF built NiTi alloy and observed that UP increases the surface hardness, reduces the surface roughness, and lessens the subsurface porosity. During UP, the material at the peaks is either removed or pushed to the valley. Besides this, unmelted particles at the sample surface will be cleared away, which increases the surface finish of the built component. The subsurface pores in an LPBF built component will either be shrunk or removed as a result of compaction during peening. In addition, the plastic strain induced during peening leads to work-hardening, which develops the surface hardness of the material. UP can also lead to nano-crystallisation of the samples, thereby increasing strength and hardness further. Corrosion resistance and wear resistance of LPBF built NiTi improved significantly after UP. In line with the surface

**TABLE 5.2**

**Mechanical Properties of LPBF-Built SS 316L Before and After UP**

| Mechanical Properties | LPBF-Built SS 316L | UP treated 316L |
|---|---|---|
| Young's modulus (GPa) | 144 | 143 |
| yield strength (MPa) | 534 | 734 |
| ultimate strength (MPa) | 630 | 862 |
| ductility (%) | 37 | 14 |

properties, an improvement in the bulk properties is also observed in LPBF built SS 316L. Table 5.2 presents the variation in mechanical properties of LPBF built SS 316L. It is reported that the yield strength and ultimate strength of LPBF built SS 316L has increased with UP. On the other hand, the ductility of the same sample reduced with UP (Joshua and Ajit 2017). A similar trend is observed for LPBF built Ti-6Al-4V in terms of yield strength, ultimate strength and ductility (Yang et al. 2018).

### 5.2.2 SURFACE PROPERTY IMPROVEMENT TECHNIQUES

#### 5.2.2.1 Laser Remelting and Polishing

Laser remelting (LR) is a technique that uses laser energy to melt the surface of a MAM-built component in order to modify its properties. During LR, quick melting and cooling take place once the laser passes away, which results in high values of cooling rates. A high cooling rate leads to finer grain structure at the surface as well as increases the surface hardness and wear resistance.

LR is used in MAM-built components to reduce the surface roughness and improve the surface properties. It can be derived from the literature that LR is generally applied to samples built using LPBF and LDED. Some of the examples on LR of MAM samples are LDED built AlCoCrFeNi based high entropy alloy (Ocelík et al. 2016), Fe-Co alloy (Yang et al. 2019) and SS 316L (Rombouts et al. 2013) alloys and LPBF built Cu-Al-Ni-Mn based shape-memory alloy (Gustmann et al. 2018), AlSi10Mg (Liu, Li, and Li 2019), Co-Cr alloy (Richter et al. 2019), SS 316L (Obeidi et al. 2019) and Ti-6Al-4V (Vaithilingam et al. 2016). Table 5.3

**TABLE 5.3**

**Surface Roughness of MAM-Built Samples Before and After Laser Remelting**

| MAM Process | Material | Roughness - Before Remelting (μm) | Roughness - After Remelting (μm) | Reference |
|---|---|---|---|---|
| LPBF | AlSi10Mg | Ra - 13.34 | Ra - 9.94 | (Liu, Li, and Li 2019) |
| LPBF | Co-Cr | Ra - 24.8 | Ra - 3.9 | (Richter et al. 2019) |
| LDED | SS 316L | Rt - 30 | Rt - 5 | (Rombouts et al. 2013) |
| LPBF | SS 316L | Ra - 10.4 | Ra - 2.7 | (Obeidi et al. 2019) |

presents the surface roughness of the samples in as-built and LR conditions for different alloys. A significant improvement in the surface finish is observed after LR irrespective of the material and MAM process. In addition, laser remelting reduces the grain size and dendritic arm spacing due to the high cooling rate. During LR, relatively lower thermal conductivity and laser absorptivity reduce the amount of energy consumed by the previous layer, and this can result in lower melt pool depth and higher cooling rate (Liu, Li, and Li 2019). Figure 5.7 presents the microstructure of LDED built samples with and without remelting. The fine grain structure developed during LR also increases surface hardness and wear resistance. An increase in the microhardness by 11% and a reduction in wear rate by 33% is reported by Yang et al. 2019 for the Fe-Co alloy. Similar observations are also seen in other alloys namely AlSi10Mg by (Liu, Li, and Li 2019). A significant reduction in porosity is also experienced after remelting due to material redistribution during laser remelting and metallurgical fusion. LR can induce a significant amount of tensile stress on the material and aid in the development of intermetallic phases, which are confirmed using X-ray diffraction by (Yang et al. 2019)

Laser polishing (LP) is an extended version of LR in which a combination of continuous-wave laser and pulsed laser-based laser melting is used to polish the metallic surfaces. LP is used to achieve a good surface finish without damaging the structural form and the geometrical features of a component. LP takes place with melting and redistribution of a thin material layer near the surface. The process is advantageous as it can be used for polishing complex shaped components that are typically built using MAM. Besides, the process does not involve tool cost and process time. LP typically shows an improvement of 90% in the surface finish (Bordatchev, Cvijanovic, and Tutunea-Fatan 2019). Figure 5.8(a)

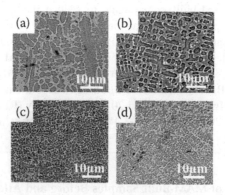

**FIGURE 5.7** Microstructure of as-built and remelted Fe-Co alloys: (a) as-built produced at a laser power of 1,400 W, (b) as-built built at a laser power of 1,600 W, (c) Built at a laser power of 1,400 W and remelted at a laser power of 2,000 W, and (d) Built at a laser power of 1,600 W and remelted at a laser power of 2,000 W. (Reprinted from *Journal of Physics and Chemistry of Solids*, 130, Yang, Xiaoshan, Jinna Liu, Xiufang Cui, Guo Jin, Zhe Liu, Yanbo Chen, and Xiangru Feng, Effect of Remelting on Microstructure and Magnetic Properties of Fe-Co-Based Alloys Produced by Laser Additive Manufacturing, 210–216, 2019, with kind permission from Elsevier).

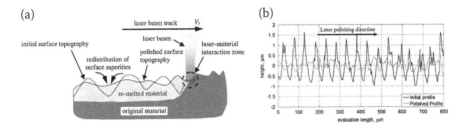

**FIGURE 5.8** Laser Polishing (a) Schematic (b) Typical Roughness Profile (Reprinted from *Applied Surface Science*, 405, Bhaduri, Debajyoti, Pavel Penchev, Afif Batal, Stefan Dimov, Sein Leung Soo, Stella Sten, Urban Harrysson, Zhenxue Zhang and Hanshan Dong, Laser Polishing of 3D Printed Mesoscale Components, 29–46, 2017, with kind permission from Elsevier).

presents a typical schematic of laser polishing. LP involves two stages: macro-polishing to smoothen the surface and micro-polishing to achieve a glossy surface finish. Macro-polishing is similar to LR discussed earlier and the depth of the remelted zone is 10–80 µm. It involves the redistribution of material due to the surface tension effects, which aids in achieving a uniform surface. Micro-polishing is performed after macro-polishing using pulsed lasers, and the depth of remelting is in the range of 5 µm. The process involves remelting a thin surface layer (< 5 µm) and the vaporisation of micro edges. The pulsed laser used for micro-polishing usually has a pulse duration of 100 ns, frequency up to 20 kHz and scanning speed greater than 0.1 m s$^{-1}$. The process parameters are selected such that molten material gets solidified before the next pulse hits the surface and creates a new one. The initial condition of the surface determines whether macro-polishing or micro-polishing has to be used or their combination is necessary. With macro-polishing roughness within the spatial wavelength from 80–1,280 µm is smoothed successfully. However, micro-polishing is necessary for the roughness with spatial wavelengths lower than 80 µm (Temmler, Willenborg, and Wissenbach 2012). Figure 5.8(b) presents the typical roughness profile of the sample with and without LP. In one of the recent investigations by (Bhaduri et al. 2017), LP is attempted using a nano-second pulsed laser and the process parameters are optimised by varying the laser fluence, percentage overlap in X and Y direction and processing atmosphere on LPBF built SS 316L. Maximum improvement in the surface finish by 94% is observed without any scratch, pits, or irregularities. Figure 5.9 presents the photographic image of polished samples at different conditions.

### 5.2.2.2 Laser Micromachining

Laser micromachining is another advanced post-processing technique used to improve the surface finish of the MAM-built part. The advantages of laser micromachining over conventional micromachining systems are the lack of tool contact, which prevents tool wear and damage while machining hard materials. The process enables the fabrication of micrometre scale features by using a pulsed

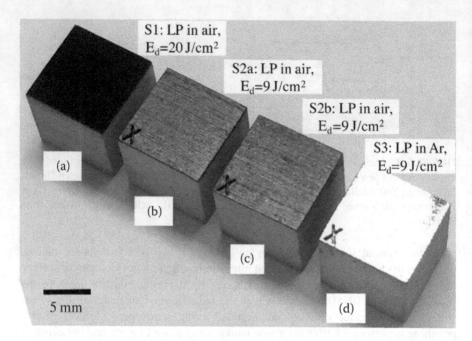

**FIGURE 5.9** Photographic view of laser polished samples (a) LP in the air at 20 J cm$^{-2}$ (b), (c) LP in the air at 9 J cm$^{-2}$ (d) LP in argon at 9 J cm$^{-2}$ (Reprinted from *Applied Surface Science*, 405, Bhaduri, Debajyoti, Pavel Penchev, Afif Batal, Stefan Dimov, Sein Leung Soo, Stella Sten, Urban Harrysson, Zhenxue Zhang, and Hanshan Dong, Laser Polishing of 3D Printed Mesoscale Components, 29–46, 2017, with kind permission from Elsevier).

laser, which deposits a fixed amount of energy into the material for precise material processing. During the laser-material interaction, material removal takes place when the energy transferred by the laser exceeds the binding energy of the material, thereby allowing cutting, scribing, drilling, or ablating the material. The mechanism behind the energy transfer is a function of the material properties and laser properties (e.g. pulse width, peak power, wavelength, etc.). Laser energy is absorbed by the material either through a thermal process or photochemical process (Liu, Du, and Mourou 1997). The thermal process is generally observed when long-wavelength lasers, like – CO$_2$ lasers, are used for machining. During the thermal process, the material is melted and/or vaporised and it is generally associated with a large heat-affected zone. The photochemical process takes place if the laser energy used for machining is greater than the bond energy of a material. In this case, bond dissociation takes place during the photon absorption, rendering thermal effects insignificant. This ablation mechanism is vital in laser micromachining. Ablation takes place when the laser energy exceeds a threshold value, and the value generally depends on the amount of energy required for ablation to take place. This, in turn, is a function of the optical and material properties of the material, like – diffusivity and absorptivity. Low diffusivity and high absorptivity will help in minimal heat dissipation and absorption of a large amount of energy, respectively. Pulse width is another major parameter in micromachining (Miller et al. 2009) (Dahotre and Harimkar 2008). Generally, femtosecond lasers are used for micromachining applications. This is because short pulse durations lead to very small heat-affected zones and provide very little time for heat dissipation. Thus, heat transfer will be constrained only to the region of interest. During micromachining, the pulse duration is kept lower than the thermal diffusion time of the material, which is in the range of 10 ps (Miller et al. 2009).

There are limited works performed on laser micromachining of MAM-built components. Worts, Jones, and Squier (2019) investigated the effect of femto-second laser-based laser micromachining of LPBF built Ti-6Al-4V. The mechanism involved in laser micromachining is plasma-mediated ablation. When the material is exposed to the laser beam, the material vaporises and gets converted to plasma directly without an intermediate liquid phase due to multiple photon absorption and subsequent ionisation. Figure 5.10(a) presents the schematic of the setup used for micromachining. The surface roughness of the as-built LPBF sample reduced from 4.23 μm to 0.8 μm in a single pass, which shows that laser micromachining is a relatively faster process (with a processing rate is 5 mm$^2$/minute) in order to achieve a high surface finish. Laser micromachining can also be used to build nano-gratings and surface micro-cones on LPBF built components. Figure 5.10(b) and 5.10(c) present the images of nano-gratings and surface micro-cones built using laser micromachining, respectively.

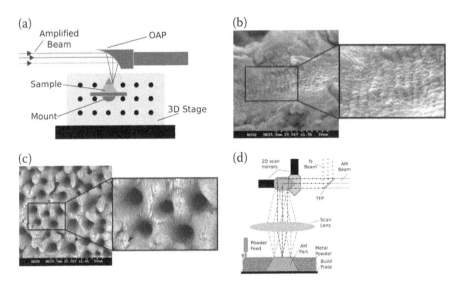

**FIGURE 5.10** Laser Micromachining of LPBF Components (a) Schematic (b) Nano-Gratings (c) Micro-Cones (d) Proposed Hybrid Stem (Reprinted from *Optics Communications*, 430, Worts, Nathan, Jason Jones and Jeff Squier, Surface Structure Modification of Additively Manufactured Titanium Components via Femtosecond Laser Micromachining, 352–357, 2019, with kind permission from Elsevier).

Femtosecond laser micromachining system can also be integrated with an existing LPBF system to have a hybrid system (refer Figure 5.10(d)) as proposed by (Worts, Jones, and Squier 2019)

### 5.2.2.3 Abrasive Finishing

Abrasive finishing is the post-processing technique that uses the flow of an abrasive medium to remove material from a metallic surface in order to achieve a high surface finish. The process is advantageous as compared to conventional finishing techniques, such as grinding and lapping, because these processes cannot be applied to highly complex components due to tool accessibility issues. Abrasive finishing can be used to achieve a high surface finish in materials that are difficult to machine, such as superalloys, ceramics, etc. The material removal takes place under three major modes – elastic deformation, plastic deformation or ploughing and micro-cutting of material. The mode of deformation is primarily a function of cutting forces, abrasive size and indentation depth generated by the abrasive (Kumar and Hiremath 2016).

Abrasive finishing is applied to MAM-built components to achieve a high surface finish at certain locations, which are difficult to access with a cutting tool. Abrasive finishing has been attempted on different materials, such as LPBF-built AlSi10Mg (Sagbas 2020) and Inconel 625 (Tan and Yeo 2017). Wang et al. (2016) reported a reduction of surface roughness of LPBF-built AlSi10Mg from 14 μm to 0.94 μm after abrasive finishing using SiC. The poor surface quality generated due to the 'balling effect' and 'powder adhesion' during LPBF is significantly removed after abrasive finishing.

Recently, ultrasonic cavitation-based abrasive finishing is done to LPBF-built Inconel 625, and it can be considered as an advanced version of abrasive finishing in which the mechanical effect due to high-intensity ultrasound is used for machining. When a high-frequency ultrasonic wave is provided in a liquid medium, cavitation will take place. During this stage, bubbles form, grow and collapse, and the disruption of the bubble generates high-velocity micro-jets. Micro-particles will act as bubble nucleation points, thus, increasing the cavitation intensity. In addition, the shock waves from the bubbles accelerate the micro-particles and increase their velocity. This results in the removal of material by micro-level ploughing and cutting. The typical ultrasonic cavitation-based abrasive finishing system consists of an ultrasonic source, transducer, horn and a container with an abrasive slurry solution having SiC particles. Figure 5.11(a) and 5.11(b) presents the improvement in surface topography with ultrasonic cavitation-based finishing. With ultrasonic cavitation-based abrasive finishing, the surface roughness of the as-built surface reduced significantly by a maximum value of 44%, as illustrated in Figure 5.11(c). 1,200 grit size produces the minimum surface finish, and it can also be noted that surface roughness of the as-built sample reduces in the absence of abrasive particles due to complete cavitation effects (Tan and Yeo 2017).

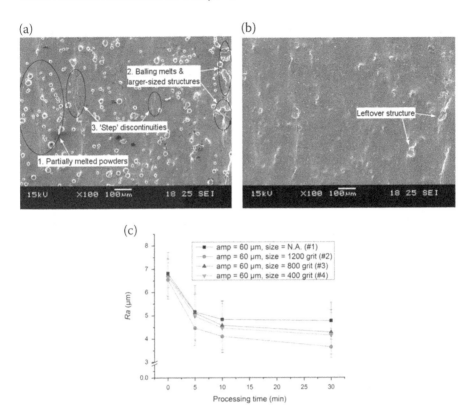

**FIGURE 5.11** Ultrasonic Cavitation-Based Abrasive Finishing (a) LPBF-built surface (b) effect of UP on surface topography (c) effect on surface roughness (Reprinted from *Wear*, 378–379, Tan, K L, and S H Yeo, Surface Modi Fi Cation of Additive Manufactured Components by Ultrasonic Cavitation Abrasive Finishing, 90–95, 2017, with kind permission from Elsevier).

## 5.3 CONCLUSIONS

This chapter has reviewed various post-processing techniques that are used for modifying the surface residual stress and improving the surface properties of MAM-built components, such as laser peening, ultrasonic peening, laser remelting and polishing, laser micromachining and abrasive finishing. Laser peening and ultrasonic peening are mainly utilised to eradicate the tensile stress on the surface of MAM-built components. However, these are also able to improve the surface properties of MAM-built components. Laser remelting and polishing, laser micro-machining and abrasive finishing are generally adopted to improve the surface properties of MAM-built components. However, laser micro-machining is also employed to generate microscopic features on the MAM-built samples.

It can be derived from the literature that even though many techniques are being attempted to post-process MAM-built samples, in-depth studies on different engineering and medical-grade alloys are required to qualify these techniques to component level. The study should also be extended towards the development of post-processing recipes for components built by different MAM techniques and material types.

## ACKNOWLEDGEMENT

A N Jinoop acknowledge the financial support from Raja Ramanna Centre for Advanced Technology, the Department of Atomic Energy, Government of India and Homi Bhabha National Institute, Mumbai, India.

## REFERENCES

Bhaduri, Debajyoti, Pavel Penchev, Afif Batal, Stefan Dimov, Sein Leung Soo, Stella Sten, Urban Harrysson, Zhenxue Zhang, and Hanshan Dong. 2017. "Laser Polishing of 3D Printed Mesoscale Components." *Applied Surface Science* 405: 29–46. doi:https://doi.org/10.1016/j.apsusc.2017.01.211.

Bordatchev, Evgueni, Srdjan Cvijanovic, and O. Remus Tutunea-Fatan. 2019. "Effect of Initial Surface Topography during Laser Polishing Process: Statistical Analysis." *Procedia Manufacturing* 34: 269–274. doi:https://doi.org/10.1016/j.promfg.201 9.06.150.

Dahotre, Narendra B., and Sandip P. Harimkar. 2008. *Laser Fabrication and Machining of Materials. Laser Fabrication and Machining of Materials.* 1st ed. New York: Springer-Verlag. doi:10.1007/978-0-387-72344-0.

Danduk, C., A. N. Jinoop, M. J. Yadav, and S. Kanmani Subbu. 2016. "Modeling and Strategies for Laser Shock Processing." *Materials Today: Proceedings* 3: 3997–4002. doi:10.1016/j.matpr.2016.11.063.

Du Plessis, Anton, Chris Broeckhoven, Ina Yadroitsava, Igor Yadroitsev, Clive H. Hands, Ravi Kunju, and Dhruv Bhate. 2019. "Beautiful and Functional: A Review of Biomimetic Design in Additive Manufacturing." *Additive Manufacturing* 27: 408–427. doi:https://doi.org/10.1016/j.addma.2019.03.033.

Du Plessis, Anton, Daniel Glaser, Heinrich Moller, Ntombizodwa Mathe, Lerato Tshabalala, Busisiwe Mfusi, and Roelf Mostert. 2019. "Pore Closure Effect of Laser Shock Peening of Additively Manufactured AlSi10Mg." *3D Printing and Additive Manufacturing* 6 (5). Mary Ann Liebert, Inc., publishers: 245–252. doi:10.1089/3dp.2 019.0064.

Galati, Manuela, and Luca Iuliano. 2018. "A Literature Review of Powder-Based Electron Beam Melting Focusing on Numerical Simulations." *Additive Manufacturing* 19: 1–20. doi:https://doi.org/10.1016/j.addma.2017.11.001.

Gibson, Ian, David W. Rosen, and Brent Stucker. 2010. *Additive Manufacturing Technologies - Rapid Prototyping to Direct Digital Manufacturing.* 1st ed. New York, NY, USA: Springer.

Gisario, Annamaria, Michele Kazarian, Filomeno Martina, and Mehrshad Mehrpouya. 2019. "Metal Additive Manufacturing in the Commercial Aviation Industry: A Review." *Journal of Manufacturing Systems* 53: 124–149. doi:https://doi.org/10.1016/j.jmsy.201 9.08.005.

Gujba, Abdullahi K., and Mamoun Medraj. 2014. *Laser Peening Process and Its Impact on Materials Properties in Comparison with Shot Peening and Ultrasonic Impact Peening.* Basel, Switzerland: Materials, MDPI. doi:10.3390/ma7127925.

Guo, Wei, Rujian Sun, Binwen Song, Ying Zhu, Fei Li, Zhigang Che, Bo Li, and Chao Guo. 2018. "Laser Shock Peening of Laser Additive Manufactured Ti6Al4V Titanium

Alloy." *Surface & Coatings Technology* 349 (June): 503–510. doi:10.1016/j.surfcoat.2 018.06.020.

Gustmann, Tobias, Holger Schwab, Uta Kühn, and Simon Pauly. 2018. "Selective Laser Remelting of an Additively Manufactured Cu-Al-Ni-Mn Shape-Memory Alloy." *Materials & Design* 153: 129–138. doi:https://doi.org/10.1016/j.matdes.2018.05 .010.

Hackel, Lloyd, Jon R. Rankin, Alexander Rubenchik, Wayne E. King, and Manyalibo Matthews. 2018. "Laser Peening: A Tool for Additive Manufacturing Post-Processing." *Additive Manufacturing* 24: 67–75. doi:https://doi.org/10.1016/j.addma.2 018.09.013.

Jinoop, A. N., S. Kanmani Subbu, C. P. Paul, and I. A. Palani. 2019. "Post-Processing of Laser Additive Manufactured Inconel 718 Using Laser Shock Peening." *International Journal of Precision Engineering and Manufacturing* 20 (9): 1621–1628. doi:10.1007/ s12541-019-00147-4.

Joshua, Gale, and Achuhan Ajit. 2017. "Application of Ultrasonic Peening during DMLS Production of 316L Stainless Steel and Its Effect on Material Behavior." *Rapid Prototyping Journal* 23 (6). Emerald Publishing Limited: 1185–1194. doi:10.1108/ RPJ-09-2016-0140.

Kalentics, Nikola, Eric Boillat, Patrice Peyre, Snežana Ćirić-Kostić, Nebojša Bogojević, and Roland E. Logé. 2017. "Tailoring Residual Stress Profile of Selective Laser Melted Parts by Laser Shock Peening." *Additive Manufacturing* 16: 90–97. doi:https:// doi.org/10.1016/j.addma.2017.05.008.

Kalentics, Nikola, Eric Boillat, Patrice Peyre, Cyril Gorny, Christoph Kenel, Christian Leinenbach, Jamasp Jhabvala, and Roland E. Logé. 2017. "3D Laser Shock Peening – A New Method for the 3D Control of Residual Stresses in Selective Laser Melting." *Materials & Design* 130: 350–356. doi:https://doi.org/10.1016/ j.matdes.2017.05.083.

Kumar, S. Santhosh, and Somashekhar S. Hiremath. 2016. "A Review on Abrasive Flow Machining (AFM)." *Procedia Technology* 25: 1297–1304. doi:https://doi.org/10.1016/ j.protcy.2016.08.224.

Liu, X., D. Du, and G. Mourou. 1997. "Laser Ablation and Micromachining with Ultrashort Laser Pulses." *IEEE Journal of Quantum Electronics* 33 (10): 1706–1716. doi:10.11 09/3.631270.

Liu, Bin, Bao-Qiang Li, and Zhonghua Li. 2019. "Selective Laser Remelting of an Additive Layer Manufacturing Process on AlSi10Mg." *Results in Physics* 12: 982–988. doi:https://doi.org/10.1016/j.rinp.2018.12.018.

Lu, Jinzhong, Haifei Lu, Xiang Xu, Jianhua Yao, Jie Cai, and Kaiyu Luo. 2020. "High-Performance Integrated Additive Manufacturing with Laser Shock Peening –Induced Microstructural Evolution and Improvement in Mechanical Properties of Ti6Al4V Alloy Components." *International Journal of Machine Tools and Manufacture* 148: 103475. doi:https://doi.org/10.1016/j.ijmachtools.2019.103475.

Liu, Shuang, Zong-Jun Tian, Xue-Song Gao, and Fei Lv. 2018. "Effect of Ultrasonic Peening on Microstructure and Properties of Laser Rapid Forming GH4169." *Optik* 172: 443–448. doi:https://doi.org/10.1016/j.ijleo.2018.07.004.

Ma, Chi, Mohsen Taheri Andani, Haifeng Qin, Narges Shayesteh Moghaddam, Hamdy Ibrahim, Ahmadreza Jahadakbar, Amirhesam Amerinatanzi, et al. 2017. "Improving Surface Finish and Wear Resistance of Additive Manufactured Nickel-Titanium by Ultrasonic Nano-Crystal Surface Modification." *Journal of Materials Processing Technology* 249: 433–440. doi:https://doi.org/10.1016/j.jmatprotec.2017.06.038.

Malaki, Massoud, and Hongtao Ding. 2015. "A Review of Ultrasonic Peening Treatment." *Materials & Design* 87. Elsevier Ltd: 1072–1086. doi:10.1016/j.matdes.2015.08.102.

Miller, Philip R., Ravi Aggarwal, Anand Doraiswamy, Yi Jen Lin, Yuan-Shin Lee, and Roger J. Narayan. 2009. "Laser Micromachining for Biomedical Applications." *JOM* 61 (9): 35–40. doi:10.1007/s11837-009-0130-7.

Obeidi, Muhannad A., Eanna McCarthy, Barry O'Connell, Inam Ul Ahad, and Dermot Brabazon. 2019. "Laser Polishing of Additive Manufactured 316L Stainless Steel Synthesized by Selective." *Materials* 12 (6): 991. doi:10.3390/ma12060991.

Ocelík, V., N. Janssen, S. N. Smith, and J. Th. M. De Hosson. 2016. "Additive Manufacturing of High-Entropy Alloys by Laser Processing." *JOM* 68 (7): 1810–1818. doi:10.1007/s11837-016-1888-z.

Paul, C. P., A. N. Jinoop, and K. S. Bindra. 2018. "Metal Additive Manufacturing Using Lasers." In *Additive Manufacturing Applications and Innovations*, edited by Rupinder Singh and J. Paulo Davium, First, 37–94. Boca Raton: CRC Press.

Richter, Brodan, Nena Blanke, Christian Werner, Niranjan D. Parab, Tao Sun, Frank Vollertsen, and Frank E. Pfefferkorn. 2019. "High-Speed X-Ray Investigation of Melt Dynamics during Continuous-Wave Laser Remelting of Selective Laser Melted Co-Cr Alloy." *CIRP Annals* 68 (1): 229–232. doi:https://doi.org/10.1016/j.cirp.201 9.04.110.

Rombouts, M., G. Maes, W. Hendrix, E. Delarbre, and F. Motmans. 2013. "Surface Finish after Laser Metal Deposition." *Physics Procedia* 41: 810–814. doi:https://doi.org/10.1 016/j.phpro.2013.03.152.

Sagbas, Binnur. 2020. "Post-Processing Effects on Surface Properties of Direct Metal Laser Sintered AlSi10Mg Parts." *Metals and Materials International* 26 (1): 143–153. doi:1 0.1007/s12540-019-00375-3.

Sames, W. J., F. A. List, S. Pannala, R. R. Dehoff, and S. S. Babu. 2016. "The Metallurgy and Processing Science of Metal Additive Manufacturing." *International Materials Reviews* 61 (5). Taylor & Francis: 315–360. doi:10.1080/09506608.2015.1116649.

Sun, Rujian, Liuhe Li, Ying Zhu, Wei Guo, Peng Peng, Baoqiang Cong, Jianfei Sun, et al. 2018. "Microstructure, Residual Stress and Tensile Properties Control of Wire-Arc Additive Manufactured 2319 Aluminum Alloy with Laser Shock Peening." *Journal of Alloys and Compounds* 747: 255–265. doi:https://doi.org/10.1016/j.jallcom.2018.02.353.

Tan, K. L., and S. H. Yeo. 2017. "Surface Modi Fi Cation of Additive Manufactured Components by Ultrasonic Cavitation Abrasive Fi Nishing." *Wear* 378–379. Elsevier: 90–95. doi:10.1016/j.wear.2017.02.030.

Temmler, A., E. Willenborg, and K. Wissenbach. 2012. "Laser Polishing." *Proceedings of SPIE - The International Society for Optical Engineering* 8243. doi:10.1117/12.906001.

Vaithilingam, Jayasheelan, Ruth D. Goodridge, Richard J. M. Hague, Steven D. R. Christie, and Steve Edmondson. 2016. "The Effect of Laser Remelting on the Surface Chemistry of Ti6al4V Components Fabricated by Selective Laser Melting." *Journal of Materials Processing Technology* 232: 1–8. doi:https://doi.org/10.1016/j.jmatprotec.2016.01.022.

Wang, Zi-meng, Yun-fei Jia, Xian-cheng Zhang, Yao Fu, and Cheng-cheng Zhang. 2019. "Effects of Different Mechanical Surface Enhancement Techniques on Surface Integrity and Fatigue Properties of Ti-6Al-4V: A Review Effects of Different Mechanical Surface Enhancement Techniques on Surface." *Critical Reviews in Solid State and Materials Sciences* 44 (6). Taylor & Francis: 445–469. doi:10.1080/1040843 6.2018.1492368.

Wang, Xuanping, Shichong Li, Youzhi Fu, and Hang Gao. 2016. "Finishing of Additively Manufactured Metal Parts by Abrasive Flow Machining." In *Proceedings of the 27th Annual International Solid Freeform Fabrication Symposium*, 2470–2472. https:// sffsymposium.engr.utexas.edu/sites/default/files/2016/197-Wang.pdf.

Worts, Nathan, Jason Jones, and Jeff Squier. 2019. "Surface Structure Modification of Additively Manufactured Titanium Components via Femtosecond Laser Micromachining." *Optics*

*Communications* 430 (July 2018). Elsevier Ltd.: 352–357. doi:10.1016/j.optcom.201 8.08.055.

Xing, Xiaodong, Xiaoming Duan, Tingting Jiang, Jiandong Wang, and Fengchun Jiang. 2019. "Ultrasonic Peening Treatment Used to Improve Stress Corrosion Resistance of AlSi10Mg Components." *Metals* 9: 103. doi:10.3390/met9010103.

Yadav, M. J., A. N. Jinoop, C. Danduk, and S. K. Subbu. 2017. "Laser Shock Processing: Process Physics, Parameters, and Applications." *Materials Today: Proceedings* 4: 7921–793. doi:10.1016/j.matpr.2017.07.128.

Yang, Xiaoshan, Jinna Liu, Xiufang Cui, Guo Jin, Zhe Liu, Yanbo Chen, and Xiangru Feng. 2019. "Effect of Remelting on Microstructure and Magnetic Properties of Fe-Co-Based Alloys Produced by Laser Additive Manufacturing." *Journal of Physics and Chemistry of Solids* 130: 210–216. doi:https://doi.org/10.1016/j.jpcs.2019.03.001.

Yang, Yichong, Xin Jin, Changmeng Liu, Muzheng Xiao, Jiping Lu, Hongli Fan, and Shuyuan Ma. 2018. "Residual Stress, Mechanical Properties, and Grain Morphology of Ti-6Al-4V Alloy Produced by Ultrasonic Impact Treatment AssistedWire and Arc Additive Manufacturing." *Metals* 8 (11): 934. doi:10.3390/met8110934.

Yin, Fei, Milan Rakita, Shan Hu, and Qingyou Han. 2017. "Overview of Ultrasonic Shot Peening." *Surface Engineering* 33 (9). Taylor & Francis: 651–666. doi:10.1080/02 670844.2017.1278838.

Zhang, Meixia, Changmeng Liu, Xuezhi Shi, Xianping Chen, Cheng Chen, Jianhua Zuo, Jiping Lu, and Shuyuan Ma. 2016. "Residual Stress, Defects and Grain Morphology of Ti-6Al-4V Alloy Produced by Ultrasonic Impact Treatment Assisted Selective Laser Melting." *Applied Sciences* 6: 304. doi:10.3390/app6110304.

# 6 Microstructure, Mechanical and Corrosion Behaviour of Cu-SiC Composites Fabricated by Stir Casting Method

Anil Kothari, Hemant Jain, Vikas Shrivastava, Richa Thakur, Pankaj Agarwal, and Sanjay Jain

## CONTENTS

## 6.1 INTRODUCTION

Copper-based composite materials are found in applications in electrical, aircraft, automobile and defence fields due to their light weight, excellent mechanical properties and corrosion performance (Hidalgo-Manrique et al. 2019; Gautam et al. 2018; Alaneme and Odoni 2016). The addition of hard reinforcement particles such as SiC (Sha et al. 2020), $Al_2O_3$ (Shi et al. 2018), CNT (Faria et al. 2018), $TiO_2$ (Bahador et al. 2020) etc. to form metal matrix composite improves mechanical characteristics and wear and corrosion properties. The various types of composite materials are aluminium-based composite, copper-based composite (Liang et al. 2019), zinc-based composite (Li et al. 2019), titanium-based composite (Cao and Liang 2020) and steel-based composite (Lee et al. 2020). However, the copper-based composite materials performed the best, because aluminium and zinc have a significantly lower melting point and are, therefore, unstable in high-temperature applications. Nevertheless, titanium and steel are high melting point materials, but these are unable to easily fabricate via melting route techniques. The copper-based composite is currently becoming an attractive candidate material for high-temperature applications. These also possess excellent wear and corrosion characteristics and adequate mechanical properties (Akramifard et al. 2014; Kumar and Parihar 2016).

Copper-based composite can be fabricated by two techniques, namely liquid melting and powder metallurgy. Among these techniques, the liquid metallurgy technique is a rather simple, easy to handle and cost-effective procedure that achieves desirable mechanical and corrosion performance (Faraji et al. 2013; Li et al. 2016). Many researchers have fabricated copper-based composite with the help of liquid melting and powder metallurgy techniques. Gan and Gu (2006) prepared copper composite using reinforcement $SiC_p$ particles (particle size 20 µm) through the powder metallurgy process. The hardness of the copper composite was reported between 60.5 to 77.6 HV, with flexural strength ranging from 215 to 203 MPa. Shabani et al. (2016) studied the microstructure and sliding wear properties of the copper composite using SiC particles in the range of 20 to 60 volume % through the powder metallurgy route. They discovered that the copper composite possesses high wear resistance and almost no wear volume as compared to pure copper. Prosviryakov 2015 studied the copper composite materials using SiC reinforcement particles in the range of 15 to 35 wt % and prepared by the hot pressing method through the powder metallurgy route. They determined that the increase in SiC particles contents of up to 25 wt % leads to greater hardness, but with SiC particles contents of over 25 wt %, the hardness lowers. Akbarpour, Salahi, Hesari, Simchi, et al. (2013) studied the processing of copper composite using the reinforcement of SiC and carbon nanotubes particles in a range of 2 to 6 wt % through the powder metallurgy route. The copper powder (size range < 20 µm) and reinforcement particles (SiC and carbon nanotubes) were mixed with the milling process. They identified that the microhardness, yield stress and elastic modulus heightens with the increased addition of reinforcement particles. Torabi and Arghavanian (2019) studied microstructure and corrosion resistance of copper-based composite (Cu-10Sn/SiC) using SiC particles in a range of 0 to 15 wt % prepared by powder metallurgy. They observed that the addition of SiC particles builds up the values of microhardness as well as the corrosion resistance of the Cu-Sn/SiC composite. Akbarpour, Salahi, Hesari, Yoon, et al. (2013) prepared copper-based composite using 2 wt % nano SiC particles (average particle size 40 nm) through the powder metallurgy process. The yield stress of 630 MPa was reported for the copper composite. Tjong and Lau (2000) investigated pure copper and copper-based composite using the reinforcement of SiC particles in the range of 0 to 20 wt % made by the hot isostatic pressing (HIP) method through the powder metallurgy route. The values of microhardness reported for the composite were between 82 to 101 HV. Furthermore, they examined that the pure copper samples resulted in higher wear loss as compared to 20 wt % composite samples. Sam and Radhika (2019) studied the mechanical and tribological properties of the copper-based composite using $Al_2O_3$ (10 wt % and average size 10 μm) produced by the horizontal centrifugal casting technique. The values of microhardness and tensile strength obtained through this were 253 HV and 289 MPa. Nageswaran, Natarajan, and Ramkumar (2018) prepared the copper composite by adding $TiO_2$ reinforcement particles in the range of 0 to 9 wt % using the stir casting method. They observed that the rising amount of $TiO_2$ reinforcement particles increases microhardness and the mechanical and wear performance of the copper composite.

The literature review indicates that most of the work on reinforcement and processing of copper composites is based on solely powder metallurgy. However, the powder metallurgy route is expensive. An alternative is to execute the process via stir casting, because this process is simple, easy and mixes particles homogenously. So far, there is a severe lack of work that has been done on stir casting-based composite development and processing. To fill the gap, the present work aims to investigate the microstructure, microhardness and corrosion properties of Cu-SiC composites processed by the stir casting method.

## 6.2 EXPERIMENTAL PROCEDURE

### 6.2.1 RAW MATERIALS

The copper composites were synthesised using commercially available pure copper plate and silicon carbide (SiC) reinforcement powder particles with 99.9% purity, and these were supplied by Sigma-Aldrich, Germany. The pure copper plates had dimensions of 50 × 50 × 5 mm, and the SiC reinforcement particles were angular and circular in shape and possess an average particle size of 35 μm.

### 6.2.2 MATERIAL PROCESSING

The copper composite was fabricated using the stir casting method through the liquid metallurgy route. The reinforcement of SiC was done in 0, 3, 6, 9 and 12 wt % in the Cu matrix. Firstly, the pure copper plates were melted at 1,150 °C using an electric muffle furnace. Second, the SiC particles were preheated at 1,200 °C with a dwell time of 2 hours. After preheating, SiC particles were added to the molten metal by using a mechanical stirrer rotating at a speed of 500 to 650 rpm for approximately five to ten minutes. The process was carried out in an argon atmosphere in order to prevent oxidation. The reinforcement particles are difficult to uniformly disperse in liquid molten metal through the mechanical stir method, so a minor modification in this method was accomplished. Finally, the molten composite material is poured into a steel die, and specimens were successively obtained. Figure 6.1 shows the schematic diagram of manufacturing Cu-SiC composite via stir casting process.

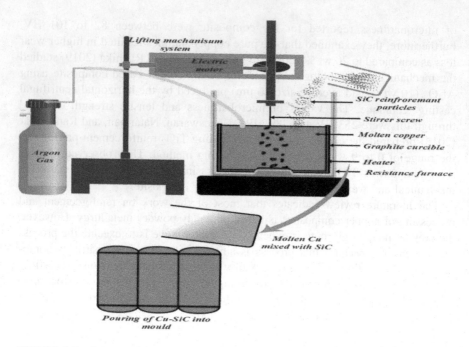

**FIGURE 6.1** Schematic Diagram of Stir Casting Experimental Step for Making Cu-SiC Composites

## 6.2.3 CHARACTERISATIONS OF COPPER COMPOSITE (CU-SIC)

After production, cylindrical Cu-SiC composite samples or specimens with an average height of 15 ± 0.5 mm and an approximate diameter of 10 ± 0.2 mm were prepared using an advanced lathe machine. The pictograph of Cu-SiC composite samples is shown in Figure 6.2. The microstructure analyses of the Cu-SiC sample were examined using a scan electron microscope. For a detailed microstructure of Cu-SiC, samples were polished and etched using chemical compositions such as 1:1; nitric acid and 99.9% pure distilled water. The polished and etched samples are cleaned with acetone using a vibration ultrasonic cleaner. The Vickers micro-hardness test was conducted on Cu-SiC samples while using a LILCA hardness tester at a load of 100 gf for 10–15 seconds. The indentations were placed at the centre of the copper composite matrix. For each case of SiC particle contents, a reading of at least 40 indentations was made to obtain the average values. Compressive tests of Cu-SiC samples were carried out using the INSTRON-8801 UTM machine at a constant strain rate 0.01 s$^{-1}$ condition and at room temperature (26°C).

## 6.2.4 CORROSION RATE

The Cu-SiC samples were immersed in 3.5% NaCl for about 30 minutes to attain an open circuit potential (OCP), and Tafel plots of Electrochemical Impedance Spectroscopy (EIS) were prepared using Corr-Ware software. Tafel plots were

**FIGURE 6.2**   SiC Reinforced Copper Composite Samples

obtained with a scan rate of 1 mV s$^{-1}$ in the range of – 0.15 to + 0.15 V relative to the OCP. The corrosion rate of the Cu-SiC sample is calculated from the resulting Tafel plot. Consequently, the corrosion rate was obtained in mpy (mils per year) (Shrivastava et al. 2017) using the following equation (Huang, Nauman, and Stanciu 2019; Sharma and Pandey 2019):

$$\textbf{Corrosion rate}(\textbf{CR}) = \textbf{0. 13} * \frac{\textbf{I}_{\textbf{corr}} * \textbf{EW}}{\textbf{D}}$$

where, $I_{corr}$ is corrosion current density in µA cm$^{-2}$, EW is the equivalent weight of the sample in g, D is the density of the sample in g cm$^{-3}$.

## 6.3   RESULTS AND DISCUSSION

### 6.3.1   Microstructure Analysis

The microstructure of copper composite sample content 9 wt % SiC particles is depicted in Figure 6.3(a). It can be observed that the SiC particles were easily and uniformly dispersed with the reinforcement particles in the copper matrix through stir casting method. Stir casting is a relatively simple and economical process being done to fabricate the composite with excellent quality. The stir casting process with a large amount of SiC particles exhibits a more uniform and homogenous distribution in the matrix rather than small particles. The SiC particles (which are marked 'black with white arrow') can be visualised at a higher magnification view in Figure 6.3(b). The SiC particles are easily incorporated into the copper matrix in

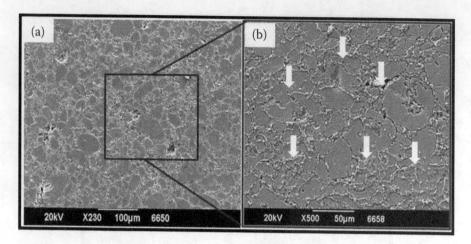

**FIGURE 6.3** (a) Microstructure of Copper Composite Samples (b) Microstructure of Cu-SiC at Higher Magnification View with SiC 9 wt %

a balanced manner. However, it was observed that the SiC particles with over 9 wt % increase the agglomeration of reinforcements particles.

## 6.3.2 MICROHARDNESS

The microhardness of the copper composite samples with varying ranges of SiC contents is shown in Figure 6.4. It is illustrated that microhardness increases with the incremental reinforcement SiC particles which is due to their higher hardness that is overall distributed in the composite, thus, making it harder.

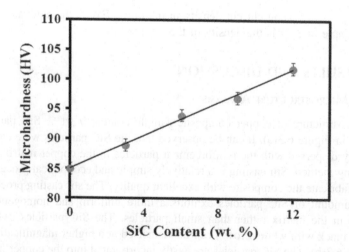

**FIGURE 6.4** Variation of Microhardness with SiC Particle Reinforcement

Mostly, SiC particles impart excellent dispersion strengthening and have adequate interfacial bonding between the copper matrix and reinforcement particles. Thus, the SiC particles enhance the microhardness of the composite. The compressive yield stress values are calculated from the compressive yield stress-strain curve. The yield stress directly indicates the strength of the materials. The compressive yield stress values of the copper composite samples with a varying range of SiC contents are illustrated in Figure 6.5. It is evident that upon increasing the SiC particles, the strength of the composites increases as well, but only up to 9 wt as it makes certain of the uniform distribution of SiC particles in the matrix. However, above 9%, the yield stress value decreases due to the agglomeration of SiC particles and the further occurrence of defects.

### 6.3.3 CORROSION BEHAVIOUR

A corrosion test was conducted on all of the samples, and the corrosion rate in mils per year (mpy) was evaluated by using $I_{corr}$. From Figure 6.6, the results showed that the corrosion rate was decreased upon increasing the reinforcement content. The minimum corrosion rate was discovered in the 9 wt % sample of about 2.6 mpy. The uniform distribution of nanoparticles makes the composite more resistant to corrosion. However, beyond 9 wt %, there was no noticed improvement in corrosion resistance. This can be attributed to the agglomeration of SiC particles, thus, the composite performance stagnates after this limit. Agglomeration leads to generate porosity inside the sample, due to which the corrosive ions penetrate the sample and, hence, corrode the sample quite critically (Shrivastava, Singh, and Singh 2017).

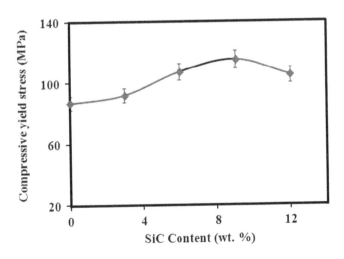

**FIGURE 6.5**  Compressive Yield Stress versus SiC Content in Cu-SiC Composites

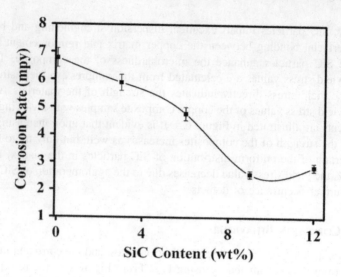

**FIGURE 6.6**  Corrosion Rate versus SiC Content in Copper Composite Samples

## 6.4  CONCLUSION

Fabrication of copper composite embedded with SiC particles by stir casting is discussed in this chapter. Based upon the investigation results on the properties of the fabricated composite, the following conclusions can be drawn from this research:

- The copper SiC composite was successfully fabricated using the stir casting method through the liquid metallurgy route.
- The reinforcements of SiC particles were achieved.
- The microhardness values were obtained in the range of 85 to 102 HV. The microhardness values were discovered to be proportionate to SiC particle percentage, i.e. an increase in microhardness values was accomplished with a corresponding increase in SiC reinforcement.
- The yield stress was attained in the range of 87 to 105 MPa and established a maximum at 9 wt % reinforcement of SiC.
- The corrosion rate was determined to decrease with a corresponding increase in SiC particle reinforcement, the least amount of which is at 9 wt %.
- Overall, the optimum content of SiC particles was identified to be 9 wt % in order to produce satisfactory mechanical properties and corrosion resistance.
- The results of the present research are found to be at par and, in some instances, better than past work conducted on the processing of Cu-SiC composites using other techniques. Therefore, the present work identifies stir casting as a viable alternative to other processes for composite fabrication.

# REFERENCES

Akbarpour, M. R., E. Salahi, F. Alikhani Hesari, A. Simchi, and H. S. Kim. 2013. Fabrication, characterization and mechanical properties of hybrid composites of copper using the nanoparticulates of SiC and carbon nanotubes. *Materials Science and Engineering: A* 572:83–90.

Akbarpour, M. R., E. Salahi, F. Alikhani Hesari, E. Y. Yoon, H. S. Kim, and A. Simchi. 2013. Microstructural development and mechanical properties of nanostructured copper reinforced with SiC nanoparticles. *Materials Science and Engineering: A* 568:33–39.

Akramifard, H. R., M. Shamanian, M. Sabbaghian, and M. Esmailzadeh. 2014. Microstructure and mechanical properties of Cu/SiC metal matrix composite fabricated via friction stir processing. *Materials & Design (1980–2015)* 54:838–844.

Alaneme, Kenneth Kanayo, and Benjamin Ufuoma Odoni. 2016. Mechanical properties, wear and corrosion behavior of copper matrix composites reinforced with steel machining chips. *Engineering Science and Technology, an International Journal* 19 (3):1593–1599.

Bahador, Abdollah, Junko Umeda, Esah Hamzah, Farazila Yusof, Xiaochun Li, and Katsuyoshi Kondoh. 2020. Synergistic strengthening mechanisms of copper matrix composites with $TiO_2$ nanoparticles. *Materials Science and Engineering: A* 772:138797.

Cao, Hong-chuan, and Yi-long Liang. 2020. The microstructures and mechanical properties of graphene-reinforced titanium matrix composites. *Journal of Alloys and Compounds* 812:152057.

Faraji, Soheila, Afidah Abdul Rahim, Norita Mohamed, Coswald Stephen Sipaut, and Bothi Raja. 2013. Corrosion resistance of electroless Cu–P and Cu–P–SiC composite coatings in 3.5% NaCl. *Arabian Journal of Chemistry* 6 (4):379–388.

Faria, Bruno, Cátia Guarda, Nuno Silvestre, José N. C. Lopes, and Diogo Galhofo. 2018. Strength and failure mechanisms of cnt-reinforced copper nanocomposite. *Composites Part B: Engineering* 145:108–120.

Gan, K. K., and M. Y. Gu. 2006. Mechanical alloying process and mechanical properties of Cu–SiCp composite. *Materials Science and Technology* 22 (8):960–964.

Gautam, Yash Kumar, Nalin Somani, Monu Kumar, and Sunil Kumar Sharma. 2018. A review on fabrication and characterization of copper metal matrix composite (CMMC). *AIP Conference Proceedings,* 020017.

Hidalgo-Manrique, Paloma, Xianzhang Lei, Ruoyu Xu, Mingyu Zhou, Ian A. Kinloch, and Robert J. Young. 2019. Copper/graphene composites: a review. *Journal of Materials Science* 54 (19):12236–12289.

Huang, Sabrina M., Eric A. Nauman, and Lia A. Stanciu. 2019. Investigation of porosity on mechanical properties, degradation and in-vitro cytotoxicity limit of Fe30Mn using space holder technique. *Materials Science and Engineering: C* 99: 1048–1057.

Kumar, Akshay, and Anubhav Singh Parihar. 2016. A review on mechanical and tribological behaviors of stir cast copper–silicon carbide matrix composites. *International Research Journal of Engineering and Technology* 3:2658–2664.

Lee, Yeong-Hwan, Namkyu Kim, Sang-Bok Lee, Yangdo Kim, Seungchan Cho, Sang-Kwan Lee, and Ilguk Jo. 2020. Microstructure and mechanical properties of lightweight TiC-steel composite prepared by liquid pressing infiltration process. *Materials Characterization* 162:110202.

Li, Zhuan, Yi-Zhong Liu, Ben-Gu Zhang, Yu-Hai Lu, Yang Li, and Peng Xiao. 2016. Microstructure and tribological characteristics of needled C/C–SiC brake composites fabricated by simultaneous infiltration of molten Si and Cu. *Tribology International* 93:220–228.

Li, Lu, Hongbo Zhang, Pin Zhou, Xianglong Meng, Lizhong Liu, Jinping Jia, and Tonghua Sun. 2019. Three dimensional ordered macroporous zinc ferrite composited silica sorbents with promotional desulfurization and regeneration activity at mid-high temperature. *Applied Surface Science* 470:177–186.

Liang, Shuhua, Weizhen Li, Yihui Jiang, Fei Cao, Gezhi Dong, and Peng Xiao. 2019. Microstructures and properties of hybrid copper matrix composites reinforced by TiB whiskers and TiB2 particles. *Journal of Alloys and Compounds* 797:589–594.

Nageswaran, G., S. Natarajan, and K. R. Ramkumar. 2018. Synthesis, structural characterization, mechanical and wear behaviour of Cu-TiO2-Gr hybrid composite through stir casting technique. *Journal of Alloys and Compounds* 768:733–741.

Prosviryakov, A. S. 2015. SiC content effect on the properties of Cu–SiC composites produced by mechanical alloying. *Journal of Alloys and Compounds* 632:707–710.

Sam, Manu, and N. Radhika. 2019. Development of functionally graded Cu–Sn–Ni/Al2O3 composite for bearing applications and investigation of its mechanical and wear behavior. *Particulate Science and Technology* 37 (2):220–231.

Sha, J. J., Z. Z. Lv, G. Z. Lin, J. X. Dai, Y. F. Zu, Y. Q. Xian, W. Zhang, D. Cui, and C. L.Yan. 2020. Synergistic strengthening of aluminum matrix composites reinforced by SiC nanoparticles and carbon fibers. *Materials Letters* 262:127024.

Shabani, Mohammadmehdi, Mohammad Hossein Paydar, Reza Zamiri, Maryam Goodarzi, and Mohammad Mohsen Moshksar. 2016. Microstructural and sliding wear behavior of SiC-particle reinforced copper matrix composites fabricated by sintering and sinter-forging processes. *Journal of Materials Research and Technology* 5 (1):5–12.

Sharma, Pawan, and Pulak M. Pandey. 2019. Corrosion rate modelling of biodegradable porous iron scaffold considering the effect of porosity and pore morphology. *Materials Science and Engineering: C* 103:109776.

Shi, Yingge, Wenge Chen, Longlong Dong, Hanyan Li, and Yongqing Fu. 2018. Enhancing copper infiltration into alumina using spark plasma sintering to achieve high performance Al2O3/Cu composites. *Ceramics International* 44 (1):57–64.

Shrivastava, V., A. Singh, and I. B. Singh. 2017. Effect of sol–gel prepared nanoalumina reinforcement content on the corrosion resistances of Al 6061-Al2O3 nanocomposite in 3.5% NaCl solution. *Materials and Corrosion* 68 (10):1099–1106.

Shrivastava, Vikas, Swati Dubey, Gaurav Kumar Gupta, and I. B. Singh. 2017. Influence of alpha nanoalumina reinforcement content on the microstructure, mechanical and corrosion properties of Al6061-Al2O3 composite. *Journal of Materials Engineering and Performance* 26 (9):4424–4433.

Tjong, S. C., and K. C. Lau. 2000. Tribological behaviour of SiC particle-reinforced copper matrix composites. *Materials Letters* 43 (5–6):274–280.

Torabi, Hossein, and Reza Arghavanian. 2019. Investigations on the corrosion resistance and microhardness of Cu–10Sn/SiC composite manufactured by powder metallurgy process. *Journal of Alloys and Compounds* 806:99–105.

# 7 Industry 4.0-Based Fault Diagnosis of a Roller Bearing System Using Wave Atom Transform and Artificial Neural Network

*Rakesh Kumar Jha and Preety D. Swami*

## CONTENTS

## 7.1  INTRODUCTION

Rolling bearings that are used in rotating machinery bear axial and radial loads. Any discrepancy occurring in the bearing element may affect machine operations and lead to machinery disintegration and breakdown over a period of time. Due to the bearing faults generated during machine operations, the failure rate in machines is 40%–50% (Henriquez et al., 2014; Rai and Upadhyay, 2016). Problems in lubrication and component mounting and severe loading are some of the reasons behind the occurrence of the pit and crack formation on the surface of the bearing element. These defective bearings not only affect machine performance but also brings about a partial to complete shutdown of the production system in the long run. The results may turn out to be catastrophic, not only in terms of economical loss because this may endanger the life of the operator as well. It is good to know the machine's health condition well in advance to avoid such failures. For this purpose, vibration-based machine condition schemes are widely adopted and employed, and these rely on the principle that in standard conditions, rotating machines acquire certain vibration patterns. Any alteration in the pattern indicates that there may be certain irregularities arising in the machine. After the time-frequency-based vibration analysis methods, the envelope spectrum analysis (Yang, 2014; Tyagi and Panigrahi, 2017) comes into the equation. This method is based on the selection of the optimum band for the demodulation of an envelope which is quite crucial. A few years later, fault diagnosis using spectral kurtosis (Antoni, 2006; Randall and Antoni, 2011) was suggested in order to utilise the short-time Fourier transform. Later methods employing cepstrum analysis (Li et al., 2009) were introduced in which the cepstrum would split the harmonics of fault frequencies over a wide frequency range for analysis. Empirical Mode Decomposition (EMD) (Han et al., 2019; Hoseinzadeh et al., 2019), Ensemble EMD (EEMD) (Yu et al., 2017; Wu et al., 2019) and wavelet transform (WT) (Wang et al., 2010) based signal analysis methods were proposed by researchers. WT (Konar and Chattopadhyay, 2011) is a multiresolution transform and decomposes the signals into different frequency sub-bands called levels and, unlike Fourier transform, it simultaneously provides time-frequency information of the signal. Signal processing techniques combined with Artificial Neural Network (ANN) (Amar et al., 2015) and deep learning techniques are the latest trend and are proving to be more effective and reliable.

Applying image processing techniques to vibration images is an emerging trend in machinery fault diagnosis. Authors, Do and Chong, 2011, proposed fault diagnosis by conversion of a vibration signal to its gray-level images and scale-invariant feature transform for fault symptom indication. Authors, Amar et al., 2015, recommended an algorithm for bearing fault diagnosis in a low noise atmosphere. The method was based on first obtaining the vibration spectrum image, cancelling the noise by 2D filtering and finally classifying the faults with the use of ANN. Authors, Khan and Kim, 2016, converted 1D vibration signals into 2D gray-level images, and faults were diagnosed by local binary pattern (LBP) of the gray image, a local histogram of LBP and a KNN- classifier.

In this paper, we have converted a 1D vibration signal into a 2D gray-level image, and the wave atom transform of the image is computed to attain its wave atom coefficients. A neural network was trained to classify faults into various categories through features extracted from the wave atom coefficients. The feature matrix is comprised of four features: Semivariance, a geostatistical parameter, statistical parameters, kurtosis and entropy and the singular values of the wave atom coefficients' matrix. In order to train the neural network, 80% data of the feature matrices are used and the remaining 20% of the data are used for testing purposes.

## 7.2  LITERATURE REVIEW

Brief literature of the related terms used in this work is described in this section.

### 7.2.1  ROLLING BEARING

Rotating types of machinery rely on rolling bearings for movement. The bearings are utilised to support axial and radial loads. Figure 7.1 presents the schematic diagram of the rolling bearing. It is comprised of an inner race (IR), an outer race (OR) and rolling elements (balls) sandwiched between the inner and the outer race.

In this chapter, we exploited the CWRU (CWRU dataset) vibration data captured on rolling bearing number 6205.

Vibrations are captured for faults at various locations of the bearings (i.e. IR, OR and ball faults) and for various fault diameters. Table 7.1 provides the specifications of bearing number 6205.

When rolling elements move over a rough surface, an impulsive wave, known as the characteristics frequency (CF) of the component, is generated. The flaws modulate this signal and produce sidebands. The spectrum thus consists of the characteristic frequency along with fault frequencies and their harmonics. The CF of a rolling element is a function of the flaw's diameter, rotating speed and geometry of the device. The CFs of a bearing element are tabulated in Table 7.2, where $d$ is the ball diameter, $D$ is the pitch diameter, $n$ is the number of balls, $f_r$ is the shaft rotation rate in Hz and $\phi$ is the contact angle.

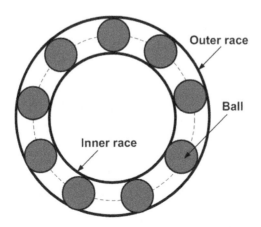

**FIGURE 7.1**  Schematic Diagram of Rolling Bearing.

## TABLE 7.1
## Specifications of Bearing Number 6205

| Parameters | Specification |
|---|---|
| outer diameter | 2.0472 inches |
| inner diameter | 0.9843 inches |
| pitch diameter | 1.537 inches |
| ball diameter | 0.3126 inches |
| shaft rotation speed | 1797 rpm |
| contact angle | $180^0$ |
| number of balls | 9 |

### 7.2.2 WAVE ATOM TRANSFORM

Wave atom transform is a multiresolution transform that adapts properly according to the texture pattern. It adheres to parabolic scaling whose period of oscillation $\lambda$ is proportional to the square of the diameter. The oscillatory nature of image textures has sparse expansion in wave atom than other transforms such as Gabor, curvelet and wavelet (Rajeesh et al., 2014). The wave atom transform captures the patterns along the oscillations and coherence of the pattern as well.

In the continuous frequency domain, it can be mathematically (Demanet and Ying, 2007) represented as presented in Equation (7.1).

$$\varphi_m^0(\omega) = exp\left(-\frac{i\omega}{2}\right)[\exp(i\alpha_m)g(\epsilon_m(\omega - \pi(m + 1/2)))$$

$$+ \exp(-i\alpha_m)g(\epsilon_{m+1}(\omega + \pi(m + 1/2)))] \qquad (7.1)$$

Where, $\alpha_m = \frac{\pi}{2}\left(m + \frac{1}{2}\right)$ and $\epsilon_m = (-1)^m$. The function g is a real-valued compact supported $C^\infty$ bump function which is selected in such a way that:

## TABLE 7.2
## Characteristic Frequencies of Bearing Components

| Characteristic Frequency | Computational Formula |
|---|---|
| cage frequency (FTF) | $FTF\ (Hz) = \frac{f_r}{2}\left(1 - \frac{d}{D}\cos\phi\right)$ |
| OR ball pass frequency (BPFO) | $BPFO\ (Hz) = \frac{nf_r}{2}\left(1 - \frac{d}{D}\cos\phi\right)$ |
| IR ball pass frequency race (BPFI) | $BPFI\ (Hz) = \frac{nf_r}{2}\left(1 + \frac{d}{D}\cos\phi\right)$ |
| ball spin frequency (BSF) | $BSF = \frac{D}{2d}\left\{1 - \left(\frac{d}{D}\cos\phi\right)^2\right\}$ |

$$\varphi_m^0(\omega) = \sum_m |\varphi_m^0(\omega)| = 1 \tag{7.2}$$

The translated wave packets $\varphi_m(t - k)$ form an orthogonal basis of $L^2(R)$. Consider 2D wave atom $\varphi_\mu(x)$ with subscript $\mu = (j, m, n) = (j, m1, m2, n1, n2)$ represented by the equation:

$$\varphi_{m,n}^j(x) = \varphi_m^j(x - 2^{-j}n) = 2^{j/2}\varphi_m^0(2^jx - n) \tag{7.3}$$

The function $\varphi_{m,n}^j(x)$ is centered in the $(x, \omega)$ space at $x_{j,n} = 2^{-j}n$ and $\omega_{j,m} = \pi2^jm$. The wave atom coefficients of $\varphi_{m,n}^j(x)$ are denoted as $c_{j,m,n}$ and can be obtained as:

$$c_{j,m,n} = \int u(x)\varphi_{m,n}^j(x)dx = \frac{1}{2\pi} \int exp(i2^{-j})n\omega\bar{\phi}_m^j(\omega)\tilde{u}(\omega)d\omega \tag{7.4}$$

In an image, curvelets provide a sparse expansion of curved edges; wavelets provide a sparse expansion of the smooth areas; and wave atoms are determined as a perfect fit for the texture present in images. For the processing of images, the best results can be obtained by judiciously employing the combination of various transforms in different areas in images so that the edges, smooth areas and texture present are correlated with the basic elements of particular transforms (Swami and Jain, 2014). $O(N)$ Wave atom coefficients are sufficient in representing an oscillatory function '$f$' to a certain extend. In contrast, we require $O(N^{3/2})$ curvelet coefficients or $O(N^2)$ wavelet or Gabor coefficients in order to represent '$f$' up to the same accuracy. Shift invariance is a significantly important property for denoising applications. The shift-invariance property of wave atoms helps in reducing denoising artefacts and, thus, ultimately owes to the absence of aliasing in the coefficients.

### 7.2.3 TEXTURE FEATURES

The coefficients of wave atom transform as defined in (7.4) are used to extract the features, which are consequently utilised to train the ANN. The features in the feature matrix consist of singular values obtained via SVD, semivariance, kurtosis and entropy of the wave atom coefficient matrix. These are further discussed in this section.

#### 7.2.3.1 Singular Value Decomposition (SVD)

Due to a large number of wave atom coefficients, the computational cost is very high. SVD can be used to reduce the size of the transformed coefficients. If $A$ is the real matrix of dimension $m \times n$, where $m \times n$, then $A$ can be represented in singular value decomposition form as:

$$A = UDV^T \tag{7.5}$$

where, the columns of matrices $U(m, n)$ and $V(n, n)$ are orthogonal, and the matrix $D(n, n)$ is diagonal with values that are positive and real known as the singular values.

The matrices U and V have orthogonal columns such that:

$$UU^T = I \tag{7.6}$$

$$VV^T = I \tag{7.7}$$

The dimensions of both of these identity matrices differ, but matrix $D$ contains only the diagonal elements.

Although SVD is mainly a dimension reduction tool, applying SVD on the transformed coefficients results in singular values that are considerably less as compared to the transformed coefficients' number. Variation of singular values of wave atom coefficients for differing textures makes it a probable candidate for texture discrimination (Rajeesh et al., 2014).

### 7.2.3.2   Semivariance

The spatial dependency between two observations can be measured using semi-variance (Pham, 2016), and is mathematically expressed as:

$$\gamma(h) = \frac{1}{2N(h)} \sum_{i=1}^{N(h)} [f(x_i) - f(x_i + h)]^2 \tag{7.8}$$

Where, $f(x_i)$ is an image intensity at a distance, $h$ $N(h)$ is the total number of $f(x_i)$ pairs separated by a distance $h$ and $f(x_i + h)$ is the intensity of the pixel $h$ distance apart from the point $x_i$.

### 7.2.3.3   Kurtosis

Kurtosis, a fourth-order central moment of data set X, is defined as:

$$\gamma_4 = \sum \frac{(X - \bar{X})^4}{nS^4} \tag{7.9}$$

Where $X$ is the observation data set, $\overline{X}$ is the mean of the observation data set, $n$ is the total number of observations in the data set and $S$ is the standard deviation.

### 7.2.3.4   Shannon Entropy

The uncertainty or disorder of presence in an event is measured by the parameter entropy and is defined as:

$$H(x) = -\sum_i p(x_i)\ln(p(x_i)) \tag{7.10}$$

Where $p(x_i)$ is the probability of an event $x_i$

### 7.2.4 ARTIFICIAL NEURAL NETWORK (ANN)

An Artificial Neural Network (ANN) is a network architecture inspired by the biological neural structure and imitates the function of the brain. Basically, it is an information processing unit that consists of nodes or neurons that are connected by links within the architecture. The links in the transmission unit have an associated weight. When information is passed through a link, it multiplies the information by weight. The output signal of a node is produced by the activation function embedded in the node.

The ANN can have one input, one output layer and one or more hidden layers. Figure 7.2 depicts the architecture of ANN structure. It possesses one input layer, one output layer and two hidden layers. An ANN with one hidden layer is called a shallow neural network, and in the case of more than one hidden layer, it is referred to as the deep neural network. If the number of hidden layers increases, then an improved learning rate is achieved, but the computational cost becomes high. The type of learning may be supervised or unsupervised depending on the problem.

## 7.3 PROPOSED METHODOLOGY

The proposed task is accomplished in five steps: (1) vibration signal acquisition, (2) gray image conversion, (3) wave atom transformation and coefficient extraction, (4) feature extraction of transformation coefficients and (5) ANN training and fault classification. Figure 7.3 shows the steps in the proposed algorithm for fault diagnosis.

### 7.3.1 VIBRATION SIGNAL ACQUISITION

For the real-time vibration signal of either a faulty and fault-free rolling bearing, the vibration data of CWRU (i.e. the CWRU dataset) data centre has been used. The fault-free and faulty rolling bearing elements (i.e. faulty inner race, faulty outer race and faulty ball) having a fault diameter of 0.007 inches sampled at a rate of 12,000 rpm

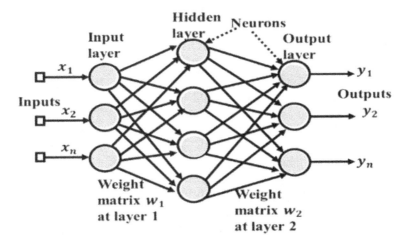

**FIGURE 7.2** Architecture of Artificial Neural Network.

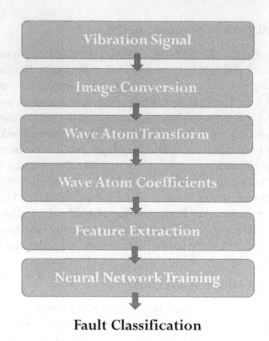

**Fault Classification**

**FIGURE 7.3**    Flow Chart of the Proposed Algorithm.

have been chosen to diagnose early-stage fault. The vibration data of rolling bearings with different load conditions (no load, 1HP, 2HP, 3HP load) and at the drive end (DE) has been tested for the abovementioned fault dimension. The details of the CWRU dataset used in this work are provided in the Appendix. The CWRU experimental setup used to capture the vibration signature of rolling bearing is depicted in Figure 7.4.

The 1D Vibration signal of healthy and three faulty conditions, i.e. inner race (IR) fault, outer race (OR) fault and ball fault (BF), of a rolling bearing for a no-load condition is shown in Figure 7.5.

### 7.3.2 VIBRATION IMAGE CONVERSION

The 2D image conversion is executed in two steps. First, a 2D vibration matrix is formed by slicing the 1D vibration signal and stacking these slices together. The vibration matrix is then converted into a gray-level image. The length of the vibration slice is determined by the sampling rate and rotational speed of the rolling bearing. For a rolling bearing rotating at a speed $S$ rpm and the data acquisition system capturing the vibration signal at a sampling rate $F_S$ Hz, the length of the slice is $l = Fs/(S/60)$ or $l = (Fs * 60)/S$. If $L$ is the length of the 1D vibration signal, then the total number of slices will be $n = L/l$. Thus, we attain a 2D vibration matrix $D(n, l)$. This vibration matrix $D(n, l)$ is converted into gray-level image $Is(n, l)$. The 2D gray-level image of the vibration signal or the vibration image of healthy and three fault conditions of a rolling bearing at a no-load, 1hp, 2hp and 3hp load condition is depicted in Figure 7.6 to Figure 7.9 respectively.

**FIGURE 7.4**   Experimental Setup for Vibration Signal Acquisition (CWRU Dataset).

### 7.3.3   FEATURE EXTRACTION

As discussed in Section 2.3, the feature matrices of images have been extracted to train the neural network. The features are derived from the wave atom coefficients of images. In this regard, we use four features extracted from the wave atom coefficients' matrix, namely: 1. singular values obtained through SVD, 2. semi-variance, 3. kurtosis and 4. entropy. Figure 7.10 shows the semivariance plot, and Figure 7.11 depicts a 3D scatter plot among SVD, kurtosis and entropy parameters of fault-free and three faulty bearing conditions at no-load.

Figure 7.10 and Figure 7.11 show that the feature parameters are clearly separable, yielding a high degree of accuracy. The same result was obtained for various load conditions, from 1hp to 3hp load. Figure 7.12 and Figure 7.13 show the semivariance and 3D scatter plot of the feature parameters at 3hp load.

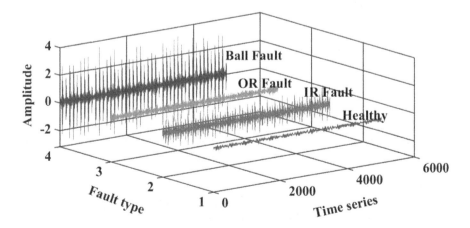

**FIGURE 7.5**   1D Vibration Signals of Healthy and Faulty Rolling Bearings.

(a)            (b)            (c)            (d)

**FIGURE 7.6**   2D Vibration Image at No Load (a) Healthy Condition, (b), (c), (d) Faulty Condition.

(a)            (b)            (c)            (d)

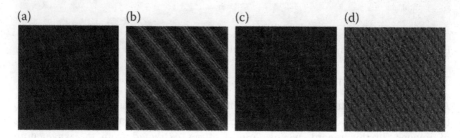

**FIGURE 7.7**   2D Vibration Image at 1hp Load (a) Healthy Condition, (b), (c), (d) Faulty Condition.

(a)            (b)            (c)            (d)

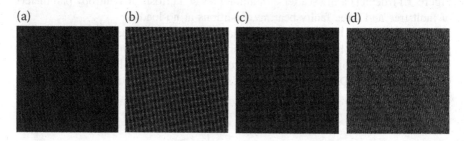

**FIGURE 7.8**   2D Vibration Image at 2hp Load (a) Healthy Condition, (b), (c), (d) Faulty Condition.

(a)            (b)            (c)            (d)

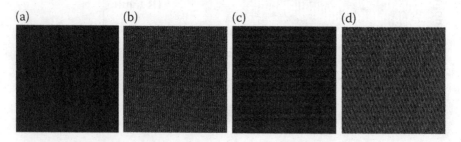

**FIGURE 7.9**   2D Vibration Image at 3hp Load (a) Healthy Condition, (b), (c), (d) Faulty Condition.

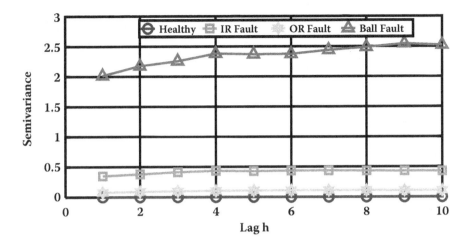

**FIGURE 7.10** Semivariogram of Healthy and Faulty Inner race, Faulty Outer race and Faulty Ball at a No-Load Condition.

### 7.3.4 FAULT DIAGNOSIS AND CLASSIFICATION

A feed-forward backpropagation neural network for fault diagnosis and classification was employed. The architecture of ANN depicts the existence of one input and one output, as well as a hidden layer with 20 neurons. The training function used is 'Trainlm', for the most advisable speed. The weights in it are updated according to the 'Levenberg-Marquardt optimization'. The size of the image is $256 \times 256$. The vibration image is subjected to five levels of wave atom decomposition. The wave atom coefficients of the fifth level have the dimensions $1248 \times 32$ and are selected for constructing the feature matrix for classification. Due to the large size of

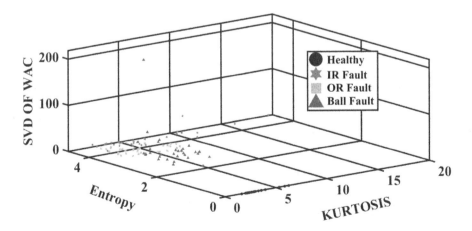

**FIGURE 7.11** 3D Scatter Plot of Singular Values of SVD, Kurtosis and Entropy of the Healthy and Faulty Inner Race, Faulty Outer Race and Faulty Ball at a No-Load Condition.

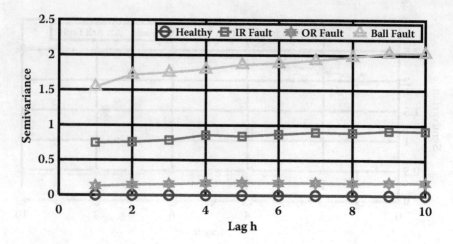

**FIGURE 7.12** Semivariance Plot of the Healthy and Faulty Inner Race, Faulty Outer Race and Faulty Ball at 3hp Load.

the coefficient matrix, the training of the ANN takes a greater computational time, so the singular values of the coefficient matrix are computed using singular value decomposition. The singular values obtained are now in a vector form of the size $32 \times 1$. Similarly, the three other features of semivariance, kurtosis and entropy of coefficients are also calculated, and each feature parameter has a size of $32 \times 1$. Combining the four features provides a feature matrix of the size $32 \times 4$, which is then fed into the four input nodes of the ANN. The ANN architecture employed in this work is shown in Figure 7.14. In order to train the ANN, 80% of the data is used while the remaining 20% data is applied in testing.

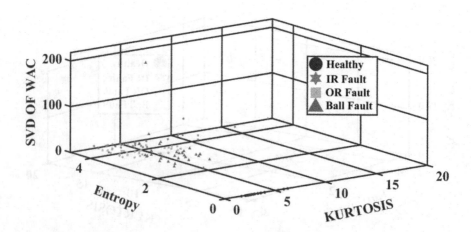

**FIGURE 7.13** 3D Scatter Plot of SVD, Kurtosis and Entropy of the Healthy and Faulty Inner Race, Faulty Outer Race and Faulty Ball at 3hp Load.

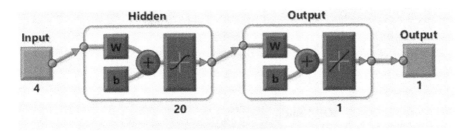

**FIGURE 7.14**   ANN Architecture for Fault Classification.

## 7.4   EXPERIMENTAL RESULTS AND DISCUSSION

From no load to 3hp load conditions, a total of four load conditions were considered in order to conduct experiments. The optimal performance of classifiers at no load and 3hp load is depicted in Figure 7.15.

For a no-load condition, the best validation performance is achieved at epoch 75, and at 3hp load condition, the best validation performance is achieved at epoch 72. Figure 7.16(a) and (b) shows the confusion matrix for no-load and 1hp load, while Figure 7.17(a) and (b) depicts the confusion matrix at 2hp and 3hp load.

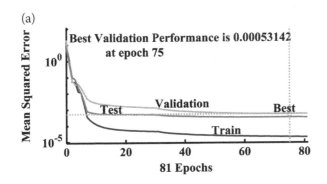

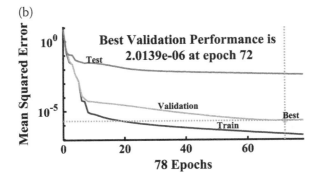

**FIGURE 7.15**   Classifier Performance Curve (a) at No-Load (b) at 3hp Load for Rolling Bearing.

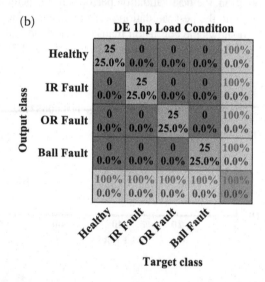

**FIGURE 7.16**  Confusion Matrices of Rolling Bearing at a Load Condition (a) 0hp (b) 1hp.

The classifier has the ability to discriminate between the healthy bearing, fault in the inner race, fault in the outer race and fault in the ball. The classification results for the various fault conditions of a rolling bearing under four different load conditions are tabulated in Table 7.3.

The classification results illustrate that the classifier can diagnose the bearing's health condition with 100% accuracy irrespective of loading condition.

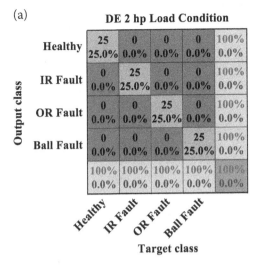

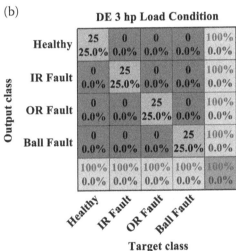

**FIGURE 7.17** Confusion Matrices of Rolling Bearing at a Load Condition (a) 2hp (b) 3hp.

## 7.5 CONCLUSION

In this book chapter, a new method of fault diagnosis that classifies the faults in rolling bearings into various categories depending upon the fault location is proposed. Vibration data is first converted into a vibration image, and wave atom transform coefficients of the vibration image are subsequently extracted. The selection of features for the formation of the feature matrix is a crucial task done for accurate classification using neural networks. Four features: singular values, semivariance, kurtosis and entropy are extracted from the wave atom transform coefficient matrix. Based on simulation results, it can also be declared that the diversity of the features selected resulted in accurate classification results. Real-time

**TABLE 7.3**

**Types of Fault and Classification Accuracies of Raw Vibration Images of Rolling Bearings**

| Fault Type | Classification Accuracy (%) | | | |
|---|---|---|---|---|
| | No Load | 1hp Load | 2hp Load | 3hp Load |
| healthy bearing | 100 | 100 | 100 | 100 |
| inner race fault | 100 | 100 | 100 | 100 |
| outer race fault | 100 | 100 | 100 | 100 |
| ball fault | 100 | 100 | 100 | 100 |
| overall accuracy | 100 | 100 | 100 | 100 |

vibration data was the basis used to verify the validity of the proposed method. Simulation results show that the proposed algorithm can efficiently locate the faults in rolling bearings. The results are also promising when the algorithm was tested with varying load conditions (i.e. 0hp, 1hp, 2hp, and 3hp). The accuracy of the classifier is 100% irrespective of load conditions, and this guarantees the universality of the algorithm.

## REFERENCES

Amar, M., I. Gondal, and C. Wilson 2015. Vibration spectrum imaging: a novel bearing fault classification approach. *IEEE Transactions on Industrial Electronics* 62(1): 494–502.
Antoni, J. 2006. The spectral kurtosis: a useful tool for characterizing non-stationary signals. *Mechanical Systems and Signal Processing* 20: 282–307.
CWRU bearing data centre, seeded fault test data, https://csegroups.case.edu/bearingdata center/pages/download-data-file.
Demanet, L., and L. Ying 2007. Wave atoms and sparsity of oscillatory patterns. *Applied and Computational Harmonic Analysis* 23: 368–387.
Do, V. T., and U.-P. Chong 2011. Signal model-based fault detection and diagnosis for induction motors using features of vibration signal in two- dimension domain. *Strojni`skivestnik— Journal of Mechanical Engineering* 57: 655–666.
Han, D., N. Zhao, and P. Shi 2019. Gear fault feature extraction and diagnosis method under different load excitation based on EMD, PSO-SVM and fractal box dimension. *Journal of Mechanical Science and Technology* 33(2): 487–494.
Henriquez, P., J. B. Alonso, M. A. Ferrer, and C. M. Travieso 2014. Review of automatic fault diagnosis systems using audio and vibration signals. *IEEE Transactions on Systems, Man, and Cybernetics: Systems* 44(5): 642–652.
Hoseinzadeh, M. S., S. E. Khadem, and M. S. Sadooghi 2019. Modifying the Hilbert-Huang transform using the nonlinear entropy-based features for early fault detection of ball bearings. *Applied Acoustics* 150: 313–324.
Khan, S. A., and J.-M. Kim 2016 Automated bearing fault diagnosis using 2D analysis of vibration acceleration signals under variable speed conditions. *Hindawi Publishing Corporation Shock and Vibration* 2016:8729572. https://doi.org/10.1155/2016/8729572.

Konar, P., and P. Chattopadhyay 2011. Bearing fault detection of induction motor using wavelet and Support Vector Machines (SVMs). *Applied Soft Computing* 11: 4203–4211.

Li, H., Y. Zhang, and H. Zheng 2009. Gear fault detection and diagnosis under speed-up condition based on order cepstrum and radial basis function neural network. *Journal of Mechanical Science and Technology* 23: 2780–2789.

Pham, T. D. 2016. *The Multiple-Point Variogram of Images for Robust Texture Classification.* In: IEEE International Conference on Acoustics Speech and Signal Processing (ICASSP-2016), Shanghai, China.

Rai, A., and S. H. Upadhyay 2016. A review on signal processing techniques utilized in fault diagnosis of rolling element bearings. *Tribology International* 96: 289–306.

Rajeesh, J., R. S. Moni, and S. S. Kumar 2014. Performance analysis of wave atom transform in texture classification. *Signal, Image and Video Processing* 8(5): 923–930.

Randall, R. B., and J. Antoni 2011. Rolling element bearing diagnostics—A tutorial. *Mechanical Systems and Signal Processing* 25: 485–520.

Swami, P. D., and A. Jain 2014. Image denoising by supervised adaptive fusion of decomposed images restored using wave atom, curvelet and wavelet transform. *Signal, Image and Video Processing* 8(3): 443–459.

Tyagi, S., and S. K. Panigrahi 2017. An improved envelope detection method using particle swarm optimisation for rolling element bearing fault diagnosis. *Journal of Computational Design and Engineering* 4(4): 305–317.

Wang, Y., Z. He, and Y. Zi 2010. Enhancement of signal denoising and multiple fault signatures detecting in rotating machinery using dual-tree complex wavelet transform. *Mechanical Systems and Signal Processing* 24: 119–137.

Wu, E. Q., J. Wang, X.-Y. Peng, P. Zheng, and R. Lob 2019. Fault diagnosis of rotating machinery using Gaussian process and EEMD-treelet. *International Journal of Adaptive Control and Signal Processing* 33(1): 52–73.

Yang, D.-M. 2014. *Induction Motor Bearing Fault Detection Using Wavelet-Based Envelope Analysis.* International Symposium on Computer Consumer and Control, Taiwan.

Yu, K., T. R. Lin, and J. Tan 2017. A bearing fault diagnosis technique based on singular values of EEMD spatial condition matrix and Gath-Geva clustering. *Applied Acoustics* 121: 33–45.

# APPENDIX

*CWRU Data Used in the Proposed Work*

**TABLE 7.A**
**12K Drive End (DE) Bearing Fault Data**

| S. No | Approx. Motor Speed (rpm) | Load (HP) | Healthy | Faulty Inner Race | Faulty Outer Race | Faulty Ball |
|---|---|---|---|---|---|---|
| 1 | 1797 | 0 | 97.mat | 105.mat | 118.mat | 130.mat |
| 2 | 1772 | 1 | 98.mat | 106.mat | 119.mat | 131.mat |
| 3 | 1750 | 2 | 99.mat | 107.mat | 120.mat | 132.mat |
| 4 | 1730 | 3 | 100.mat | 108.mat | 121.mat | 133.mat |

# 8 Advanced Sheet Metal Forming Using Finite Element Analysis

*Gajendra Kumar Nhaichaniya and Chandra Pal Singh*

## CONTENTS

## 8.1  INTRODUCTION

Forming is one of the most popular manufacturing processes done to produce complex geometries and is especially used in automobile components, automobile body parts and aerospace industrial parts. Single-stage and multi-stage forming are common processes, because these are particularly simple and only the initial tool design incurs cost. In single-stage forming, a single punch converts the blank into the desired shape, while in multi-stage forming, two or more strokes of the punch convert the blank into the required shape. In the process of forming, there are many design parameters that need to be addressed. The design of punch, die, blank holder and other tools are the machine parameters, while blank holding force or binder force, friction coefficient, punch pressure and punch velocity are the important process parameters. Material properties and initial blank shapes are also pertinent parameters. The optimum values of these parameters must be set in order to attain a defect-free output. Wrinkle at the flange and cup wall, tearing at bottom of the cup and regions of sharp angles are the main errors that are generally found in formed products (Cwiekala et al. 2011). Wrinkling can be easily controlled by blank holding force, and simulation tools can predict accurate required force (Großmann et al. 2012). In the case of a thin blank, the coefficient of friction plays a vital role in minimising wrinkling (Karupannasamy et al. 2012; Singh et al. 2017).

Uniform thickness distribution and extending of accuracy are concerns in the forming process. Geometrical accuracy and thickness distribution of the formed product are more uniform in multi-stage forming (Gonzalez et al. 2019). The digital image correlation technique is used to carry out a comparison between single-stage and multi-stage forming while considering the thickness of the part as the main criterion (Gonzalez et al. 2019).

Thin-walled geometries present in aeronautics are successfully formed by a two-stage forming process, which otherwise would not be possible by single-stage forming (Yao et al. 2018). In such thin geometries, the strain flow direction is more important in order to convert the optimum blank size into desired shape (Yao et al. 2018). Also, strain rate has an effect both on forming and on the quality of surface finish (Sigvant et al. 2019).

There are industrial requirements necessary to form a product at a large drawing depth. Long cups of titanium alloy with improved surfaces are formed by multi-stage forming (Harada et al. 2014). In multi-stage forming, a product can be heat-treated to achieve an improved surface finish. Multi-stage friction stir forming enhances form-ability, as compared with single-stage forming, and better forming limits with height-ened wall angle and increased elongation in the meridian direction as a result of circumferential strain could be therefore achieved (Otsu et al. 2017). The circumferential strain is greater than that in a single stage, and this shows that during forming, stretching took place in the circumferential direction, along with shearing. Moreover, multi-stage friction stirs forming boosts formability compared with single-stage forming (Otsu et al. 2017). Biaxial stretching exists in multi-stage incremental forming rather than plane strain condition that exists in single-stage forming, and stretching keeps on improving as the number of stages increases. This phenomenon of multi-stage forming enables the formation of a conical cup (Suresh et al. 2018).

The vertical walled product cannot be formed by single-stage forming due to excessive thinning at the transition region. Rather, up to a certain point, the vertical and angled wall product can be produced by single-stage forming (Vignesh et al. 2020). Multi-stage incremental forming can produce a vertical walled cup. An improved forming limit diagram, better strain path and enhanced forming path are achieved in multi-stage incremental forming.

The multi-stage forming process design is a particularly complex task. In each stage, deformation will definitely happen, and each previous stage affects the next successive stage. Based on strain distribution, the amount of deformation in each stage is calculated, and a minimum stage with maximum deformation is the main consideration when designing multi-stage forming (He et al. 2010; Zabaras et al. 2003). Forming tool design is the foremost important task in forming, because accurate tool designing could reduce the number of trials needed to reach the re-quired shape; however, the design of forming tool for complex components mainly depends on the engineer's experience (Chen et al. 2007). FE analysis can be used to reduce the trial time and cost of production based on virtual simulation results and plots. During the development process of the tool, the design can be easily modified and finalised using the software. This is considerably more efficient compared to the traditional approach (Guo et al. 2018).

Tool setup and the number of stages in sheet metal work mainly depend on the rule-of-thumb height to diameter ratio (h/d) and the complexity of the component. As the h/d ratio increases in the deep drawing process, the number of stages also increases (Abdelmaguid et al. 2013). For complicated structures, where the flow material is restricted in critical zones, ensuring the quality of the final component of multi-stage forming is recommended, even for a lower h/d ratio.

However, if possible, the multi-stage forming process generally avoids this because of high tooling cost, and it decreases the production rate as well. The needs of a multi-stage process in improving formability and reducing the thinning percentage within 25% in the final product are the main foci of the study, together with a comparison with single-stage forming. The HyperForm 14.0 version is used for simulating an inbuilt radio solver.

## 8.2  TOOL SETUP

The brake plate component is formed with single-stage and multi-stage forming processes. All press tool standards and clearance were considered in the design and analysis stages. Tool parameters for both forming processes were kept the same, except for punch travel height because of different blank shape and position. In the first stage for the multi-stage forming process, punch travel height and shape need to be optimised in such a manner that these will provide the desired output in the second stage. For this to happen, the requirement is not only engineering experience, but also the number of trials. In the forming tool setup, the punch, die, blank holder and blank components are shown in blue, red, green and grey respectively in the succeeding figures.

### 8.2.1  TOOL SETUP FOR THE SINGLE-STAGE FORMING PROCESS

The circular metal plate (blank) is first formed with the single-stage process. In this procedure, the blank is directly placed over the die, facing with necessary supporting items. Once the punch starts deforming the blank plate, these support structures help to keep the blank in position. The tool setup for the single-stage forming process is shown in Figure 8.1.

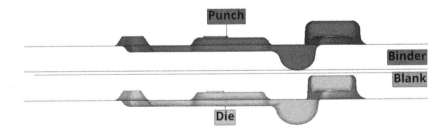

**FIGURE 8.1**   Single-Stage Forming Tool Setup.

## 8.2.2  Tool Setup for the Multi-Stage Forming Process

In the multi-stage forming process, the blank is formed in two stages. To attain optimum results in the second stage, different trials are made by changing the first stage punch shape and travel height. The first trial considered is with the cup-shaped punch and travel height of 30 mm. By general observation of the first trial's results, it is clear that a dome shape is more advisable instead of a full cup shape to achieve better formability, as shown in Figure 8.2. Moreover, other trials with different travel heights are listed in Figure 8.3. To lower overall tooling cost, the open die is used in the first stage.

As shown in Figure 8.3, the tool setup focuses on the forming stages only. To be able to finalise the manufacturing of the brake plate, more stages are required. Other process parameters are listed in Table 8.1.

## 8.3  FE ANALYSIS

In FE analysis. only the punch, die, blank holder, support and initial blank component's geometry need to be imported into simulation software. To reduce the complexity of FE analysis, only the blank component is considered as the deformable part, and the tool components (i.e. punch, die, blank holder) are considered as rigid parts, with the assumption that tool material is sufficiently hard compared to blank material. This assumption is valid because the analysis mainly focuses on the manufacturing process of the blank component rather than tool life.

Once the geometry is imported, cleanup work may be required in order to overcome the issue of data loss. The blank component is meshed by two-dimensional four-node shell elements (SHELL4N). In the property, the SH_ORTH card was selected in order to define the orthotropic shell behaviour of elements. This consists of major parameters such as Ishell formulation and the number of integration points (N) and thickness (t). Ishell 24 is selected for quadrilateral elastoplastic physical hourglass controlled (QEPH) shell element formulation. QEPH formulation of shell elements is more accurate for elastic or elastoplastic materials, irrespective of load type. Parameter t is equal to the blank thickness, and $N = 5$ is generally recommended in order to precisely capture the orthotropic plastic behaviour of the material. Tool components are meshed with two-dimensional four-node shell elements, but to be able to make these undeformable, one rigid element must be attached to each component respectively. The total generated elements for tool and blank components are 55,956 and 1,012 respectively. However, the tool component's elements do not bear individual DOF, because their DOF depends on the rigid element associated with them. This significantly reduces the overall DOF of the system; thus, computational time also decreases drastically. At this point, the overall computational time consumed by the explicit solver depends on the total number of elements in the blank component and critical time step size. If critical time step size improves, then a larger stable time increment can be considered by the explicit solver. Hence, the total number of increments reduces. The critical time

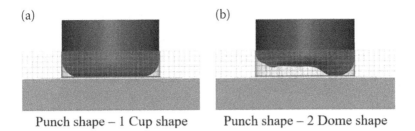

Punch shape – 1 Cup shape        Punch shape – 2 Dome shape

FIGURE 8.2   First Stage Different Punch Shapes for Forming.

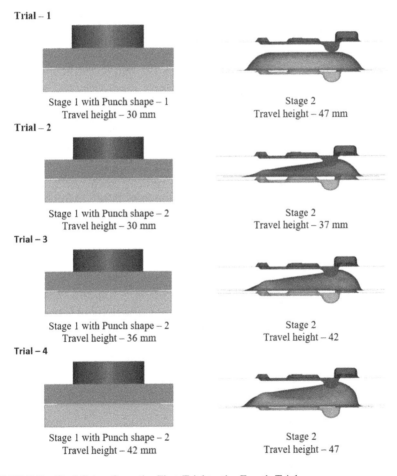

Trial – 1

Stage 1 with Punch shape – 1          Stage 2
Travel height – 30 mm           Travel height – 47 mm

Trial – 2

Stage 1 with Punch shape – 2          Stage 2
Travel height – 30 mm           Travel height – 37 mm

Trial – 3

Stage 1 with Punch shape – 2          Stage 2
Travel height – 36 mm           Travel height – 42

Trial – 4

Stage 1 with Punch shape – 2          Stage 2
Travel height – 42 mm           Travel height – 47

FIGURE 8.3   Tool Setup from the First Trial to the Fourth Trial.

**TABLE 8.1**

**Process Parameters**

| Tool Setup | Blank Material | Blank Thickness | Punch Travel Height | Punch Speed | Blank Holder Force |
|---|---|---|---|---|---|
| single-stage | CRDQ steel | 2.5 mm | as per trial number | 2000 mm/sec | 40 KN |
| multi-stage, stage I | CRDQ steel | 2.5 mm | as per trial number | 2000 mm/sec | 40 KN |
| multi-stage, stage II | CRDQ steel | import from 1st stage | as per trial number | 2000 mm/sec | 40 KN |

step size depends on material parameters and element size. To reduce computational time further, initially coarse mesh assigned to the blank component improves stable time increment size, but small geometric features may not be captured in the same element size. To overcome this issue, the adaptive mesh technique with a sub-divisions option is implemented. In this study, element size ten is used for the blank component, with adaptive subdivisions at level 2 and an angle criterion of 5 degrees.

In a sheet metal forming process, the inertia effect is not dominant; hence, the kinetic energy (KE) of the system is negligible compared to internal energy. To improve computational efficiency further, advanced mass scaling (AMS) is used to increase the elementary time step in the explicit solver, in case the time step is less than the targeted value. The technique of AMS is similar to traditional mass scaling; however, in AMS, artificially added mass does not elevate the total kinetic energy (KE) of the system. It is recommended that the kinetic energy of the system should not exceed 5% of internal energy when applying mass scaling.

An orthotropic cold rolled steel material model (e.g. hill orthotropic material model) is considered with the blank component including Lankford coefficients r0, r45 and r90. The interaction among tool components with the blank is shown in Table 8.2. There is no direct tool interaction between tool components.

**TABLE 8.2**

**Contact Definition**

| Contact Number | Component 1 | Component 2 | Coefficient of Friction |
|---|---|---|---|
| 01 | punch | blank | 0.125 |
| 02 | die | blank | 0.125 |
| 03 | blank holder | blank | 0.125 |

## 8.4 SIMULATION RESULTS AND DISCUSSION

### 8.4.1 THINNING

The measurement of the reduction in sheet thickness is the simplest way that makes thinning criteria useful in forming industries. It is broadly considered a failure criterion. If the thinning percentage exceeds the limiting value, then it is considered a failure. Nevertheless, different studies recommended against considering thinning as a failure criterion because of its variable nature and its dependence on the strain path.

In the sheet metal forming process, by the calculated amount of the blank holder force, material flow can be controlled in such a manner that only an acceptable amount of thinning and wrinkle will appear. In other words, by varying the blank holder force, wrinkling and thinning can be controlled. However, in the case of complex-forming for the same blank holder force, materials have a tendency to become stretched instead of flowing in the die cavity. Due to excessive stretching of the material, tearing happens in the blank. If the blank holder force decreased, then a lot of wrinkling appears.

From the simulation results presented in Figure 8.4, it is clear that, in the single-stage forming, a high value of thinning enhances the probability of tearing in the material on many regions. Nonetheless, the material is sufficiently ductile enough to fulfil the requirement, but due to the lack of flow of material in complex zones, localised thinning and tearing appear. The bidirectional flow of material at critical zones is one of the reasons for stretching in the sheet.

In the single-stage output, the maximum thinning is 60% in region 1, and the maximum thickening is 8%, as shown in Figure 8.4. Such a high thinning percentage does not fit the criteria; hence, the forming process must be improved. To avoid the lack of flow of material in complicated regions, the blank first is formed in a pre-calculated shape and then reformed in the second stage. In this manner, excessive stretching of material can be avoided by the Bauschinger effect. When stresses are applied in the reverse direction, as applied in stage 2, the dislocations are now aided by the back stresses that were present in the previous forming stage 1.

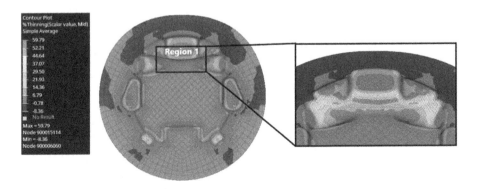

**FIGURE 8.4**   Thinning and Thickening Percentage in Single-Stage Forming.

Accordingly, the dislocations in the forming stage 2 are able to glide more easily, resulting in lower yield stress for plastic deformation for the reversed direction of loading or, in other words, smooth material flow is achieved.

In Figure 8.5, the first, fifth and tenth frames, out of 11 frames, of the single-stage and multi-stage process illustrated. For step-by-step comparison, stage 2 of the multi-stage process is considered. The first frame shows the placement of the blank over the die face. In single-stage forming and due to the abrupt change in die face, the blank material is restricted to flowing in complex regions, as shown in the fifth frame. However, in the multi-stage forming, this problem can be sorted out by considering the pre-deformed shape. The tenth frame provides a clear view of excessive stretching at a marked location in the single-stage forming process, but in the multi-stage forming process, material smoothly flows in the die cavity in order to obtain the required shape.

For Trial-1 Stage 1, maximum thinning is at 03.39%. Once the simulation of Stage 1 finishes the formed shape, the results are imported as blank in Stage 2 in order to proceed with the simulation further. Maximum thinning and thickening in Stage 2 are 26.62% and 20.49% respectively. Subsequent results for different trials made in the multi-stage forming process are shown in Figure 8.6.

Trial-4 provides thinning results within the required criteria, which is less than 20%. Thickening is more when compared to other trials, but within the acceptable range at locations. The further increase in punch travel height does not significantly reduce the thinning percentage but instead increases wrinkling which is not acceptable. However, if increasing punch travel height does not generate results with the acceptable criteria, then tool design needs improvement.

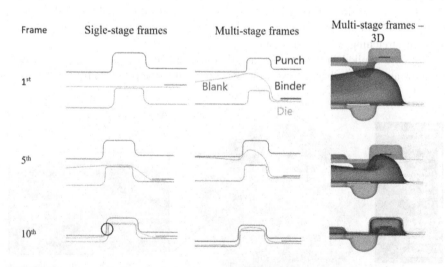

**FIGURE 8.5**   Material Flow in Single-Stage and Multi-Stage Forming Process.

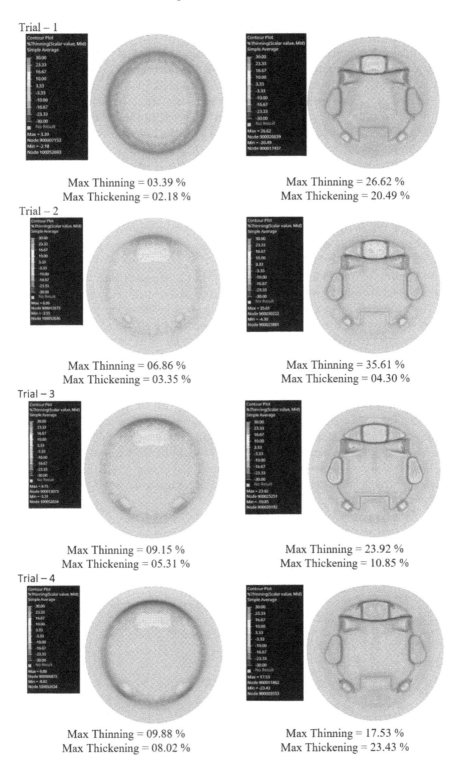

Trial – 1

Max Thinning = 03.39 %           Max Thinning = 26.62 %
Max Thickening = 02.18 %         Max Thickening = 20.49 %

Trial – 2

Max Thinning = 06.86 %           Max Thinning = 35.61 %
Max Thickening = 03.35 %         Max Thickening = 04.30 %

Trial – 3

Max Thinning = 09.15 %           Max Thinning = 23.92 %
Max Thickening = 05.31 %         Max Thickening = 10.85 %

Trial – 4

Max Thinning = 09.88 %           Max Thinning = 17.53 %
Max Thickening = 08.02 %         Max Thickening = 23.43 %

FIGURE 8.6    Thinning and Thickening Percentages in Multi-Stage Forming.

## 8.4.2 Forming Limit Diagram (FLD)

Due to extensive tensile deformation at the localised area, necking and tearing are common in many sheet forming operations. At the same time, a large compressive force is responsible for buckling in the sheet. Such complex loading is not possible to be analysed by a simple uniaxial tensile test. The uniaxial tension test has its limitations when dealing with formability in sheet metals. The state of stress during sheet metal forming is triaxial in nature, evaluating it with the uniaxial tension test will generate misleading results. A well-known forming limit diagram (FLD) is used to investigate such complex loading conditions in sheet metal forming. In software, FLD is plotted by measuring strain in two principal directions (i.e. principal strains) in the circle grid for each element. Principal strain can be calculated as the percentage change in the length of the major and minor axes. The plot of the major strain versus minor strain provides the limiting strains corresponding to deformations. A combination of major and minor strain represents failure elements (by tearing) above the forming limit curves (FLC) and safe elements below the FLC.

In Figures 8.7 and 8.8, the limiting line (FLC), marginal line, shear line and major-minor lines are illustrated. Elements between the marginal line and shear line are marked as safe elements, and elements below the shear line marked as compression elements where thickening may appear. Elements with minimum or no strain are marked in loose material at the centre of the FLD. It is observed that the slope of FLC decreases while increasing the strain hardening exponent, building up the thickness of the blank resulted in increase in FLD.

For the single-stage tool setup, a huge number of elements is above the FLC and is representing the failure element as explained in Figure 8.7. Such high strain occurs on localised elements due to excessive stretching of the material. No element suffers from high wrinkling. Both high thinning and high wrinkling conditions are undesirable, and these can be accepted up to certain limiting values only.

Trial-4 of the multi-stage process produced favourable results because of the change in strain path. In Figure 8.8, all elements are under the safe forming zone or within an acceptable range.

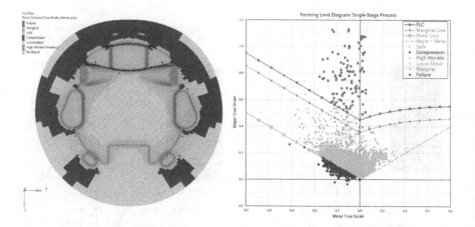

**FIGURE 8.7**   Forming Limit Diagram of the Brake Plate in the Single-Stage Process.

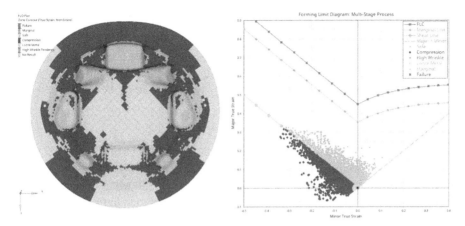

**FIGURE 8.8**   Forming Limit Diagram of the Brake Plate for the Multi-Stage Process.

## 8.5   VALIDATION OF SIMULATION RESULTS

From the acceptable range of FE analysis results, the tool design finalised with manufacturing standards, and the brake plate is manufactured to validate the results. The regions marked in Figure 8.9 are the main observation points for thinning measurements. Subsequently, the manufactured component is cut into pieces along the vertical and horizontal lines, as shown in Figure 8.9. Next, the thickness is

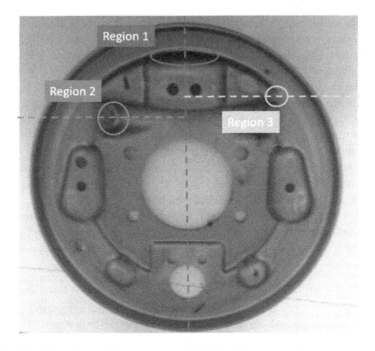

**FIGURE 8.9**   Manufactured Component for Thinning Measurement.

measured on the marked regions with the help of a screw gauge. The percentage reduction in sheet thickness is measured as ~20% by screw gauge and 17.5% in the simulation result. This means simulation and experimentation results have a high level of precision with 2.5% error only. Due to the presence of high thinning and tearing in the single-stage simulation result, this outcome does not need to be validated.

## 8.6 CONCLUSION

In single-stage forming, the maximum developed Von Mises Stress Criterion is 661.8 MPa, which is beyond the ultimate strength (i.e. 550 MPa) of the material. Such a high value of stress is enough to tear the material during the forming process. However, in the multi-stage process, lower yield stress for plastic deformation can be achieved by reversing the direction of loading. The maximum developed stress in complex regions of the multi-stage forming process is 474 MPa, which is lower than the ultimate strength of the material but sufficient enough to deform the material plastically. However, the decision of selecting the number of stages in the forming process for complex structures still depends on tool design experience.

In the formation of the brake plate where the single-stage process fails to give the desired output, multi-stage forming is highly suitable. Certainly, the multi-stage process also consists of a significant amount of thinning and wrinkles, but this is within the acceptable range. The admissible range of these parameters varies with application, quality and industrial standards. Often, the number of necessary trial iterations is identified by changing the design and/or process parameters.

In this study, iterations are made by varying the punch travel height of Stage 1 from 30 mm to 45 mm, with an increment of 3 mm. The optimum results were found at 42 mm. In this iteration, the thinning was minimum at approximately 17.5%. The thickening is 23.43% which is also within the acceptable range. The further increase of the punch travel height of Stage 1 does not provide a compelling amount of significant reduction in thinning, but thickening increases beyond the acceptable range.

The component manufactured by the multi-stage process is identified to have approximately the same type of thinning as provided by FE results, which is within 2.5% of error. Hence, this validates the results of the simulation. Complex sheet structure needs to form in two or more stages, but this is not always the case. It may possible for some instances to achieve desired outputs in a single stage only by changing tool assembly, providing additional supports and using better lubrication. Such changes may increase tool complexity, thereby increasing costs as well.

## REFERENCES

Abdelmaguid, T. F., R. K. Abdel-Magied, M. Shazly, and S. Wifi Abdalla 2013. A dynamic programming approach for minimizing the number of drawing stages and heat treatments in cylindrical shell multistage deep drawing. *Computers & Industrial Engineering* 66 (3): 525–532.

Chen, W., Z. J. Liu, B. Hou, and R. X. Du 2007. Study on multi-stage sheet metal forming for automobile structure-pieces. *Journal of Materials Processing Technology* 187: 113–117.

Cwiekala, T., A. Brosius, and A. E. Tekkaya 2011. Accurate deep drawing simulation by combining analytical approaches. *International Journal of Mechanical Sciences* 53 (5): 374–386.

Gonzalez, M. M., N. A. Lutes, J. D. Fischer, M. R. Woodside, D. A. Bristow, and R. G. Landers 2019. Analysis of geometric accuracy and thickness reduction in multistage incremental sheet forming using digital image correlation. *Procedia Manufacturing* 34: 950–960.

Großmann, K., L. Penter, A. Hardtmann, J. Weber, and H. Lohse 2012. FEA of deep drawing with dynamic interactions between die cushion and process enables realistic blank holder force predictions. *Archives of Civil and Mechanical Engineering* 12: 273–278.

Guo, Z., P. Lasne, N. Saunders, and J. P. Schillé 2018. Introduction of materials modelling into metal forming simulation. *Procedia Manufacturing* 15: 372–380.

Harada, Y., Y. Maeda, M. Ueyama, and I. Fukuda 2014. Improvement of formability for multistage deep drawing of Ti-15V-3Cr-3Sn-3Al alloy sheet. *Procedia Engineering* 81: 819–824.

He, D. H., X. Q. Li, D. S. Li, and W. J. Yang 2010. Process design for multi-stage stretch forming of aluminium alloy aircraft skin. *Transactions of Nonferrous Metals Society of China* 20 (6): 1053–1058.

Karupannasamy, D. K., J. Hol, M. B. de Rooij, T. Meinders, and D. J. Schipper 2012. Modelling mixed lubrication for deep drawing processes. *Wear* 294: 296–304.

Otsu, M., T. Ogawa, T. Muranaka, H. Yoshimura, and R. Matsumoto 2017. Improvement of forming limit and accuracy in friction stir incremental forming with multistage forming. *Procedia Engineering* 207: 807–812.

Sigvant, M., J. Pilthammar, J. Hol, J. H. Wiebenga, T. Chezan, B. Carleer, and T. van den Boogaard 2019. Friction in sheet metal forming: Influence of surface roughness and strain rate on sheet metal forming simulation results. *Procedia Manufacturing* 29: 512–519.

Singh, C. P., and G. Agnihotri 2017. Formability analysis at different friction conditions in axis-symmetric deep drawing process. *Materials Today: Proceedings* 4, (2): 2411–2418. Elsevier.

Suresh, K., S. P. Regalla, and N. Kotkundae, 2018. Finite element simulations of multi stage incremental forming process. *Materials Today: Proceedings* 5 (2): 3802–3810.

Vignesh, G., C. Pandivelan, and C. S. Narayanan, 2020. Review on multi-stage incremental forming process to form vertical walled cup. *Materials Today: Proceedings* 27: 2297–2302.

Yao, W. A. N. G., L. A. N. G. Lihui, E. Sherkatghanad, K. B. Nielsen, and C. Zhang 2018. Design of an innovative multi-stage forming process for a complex aeronautical thin-walled part with very small radii. *Chinese Journal of Aeronautics* 31 (11): 2165–2175.

Zabaras, N., S. Ganapathysubramanian, and Q. Li 2003. A continuum sensitivity method for the design of multi-stage metal forming processes. *International Journal of Mechanical Sciences* 45 (2): 325–358.

# 9 An Overview of Human Bone, Biomaterials and Implant Manufacturing

*Pradeep Singh, Pankaj Agarwal, I.B. Singh, and D.P. Mondal*

## CONTENTS

## 9.1 INTRODUCTION

Bones provide the structure to the body by the formation of a skeleton. Besides providing a framework, these also work as a substrate for combining the muscles. A bone also houses the different biological compounds that are essential for creating the new blood cells in the body (hematopoiesis) and maintaining stable equilibrium in the bodily fluid by storing ions of calcium and phosphate (i.e. homoeostasis) (Wang, Leng, and Gong 2018). While small injury or defects in the bone can be self-repaired, large defects in the bone due to severe accidents, cancerous bone or chronic bone disease cannot be healed by itself; therefore, bone grafting, bone scaffold or bone implantation is required. A bone implant or scaffold is a three-dimensional matrix made of biomaterials that mimic the natural bone and promotes the newborn bone tissues to grow on its surface.

The increase in the number of vehicle accidents and sports-related injuries, the rapid population growth of the elderly and disease related to the bone tissues have given rise to the market demand for bone implant industries all over the world. It is expected that the global market for this may increase at a compound annual growth rate of 6.8% for the forecast period 2018–2023. The global market value of a bone implant in May 2018 was about US$375 million. However, due to expensive bone implant products, instability and corrosion of the bio-implant and the government policy may hinder the market growth of the bone implant industries.

Section 9.2 of this chapter describes composition of bone, its structure, regeneration and properties. Section 9.3 sheds light on bone implantation and tissue engineering, whereas Section 9.4 provides details on bone implant materials. A comprehensive digest on bone implant synthesization is provided in Section 9.5 Finally, Section 9.6 summarises and recommends future research directions in this area.

## 9.2 HUMAN BONE COMPOSITION, REGENERATION AND PROPERTIES

### 9.2.1 Composition and Structure

The internal part of bones contains calcium and minerals required for the functioning of nerves and muscle cells (Del Valle et al. 2011). Besides these, bone is composed of carbonate, phosphorous, sialic acid and hydroxyproline as listed in Table 9.1. The percentage amount of these compounds in the largest bone of the human body, which is called the femur or thighbone, is broken down in this table (Quelch et al. 1983). However, the content of the observed minerals and compounds depends on the age, lifestyle and health of the human being in question.

The hierarchical structure of a bone is shown in Figure 9.1(a) (Sadat-Shojai et al. 2013). As observed in this figure, the outermost layer of the bone is the combination of vascular tissues that covers the bone and is known as the periosteum. After the

**TABLE 9.1**
**Different Minerals and Compounds found in the Femur of the Adult Human**

| Compound | Composition |
| --- | --- |
| mineral | 66.8% |
| organic compounds | 27.5% |
| carbonate | 5.6% |
| non-collagenous organic matrix | 2.8% |
| calcium | 258.7 mg/g |
| phosphorous | 116 mg/g |
| Ca/P | 2.24 mg/mg |
| sialic acid | 1.96 μmol/g |
| hydroxyproline | 32.8 mg/g |

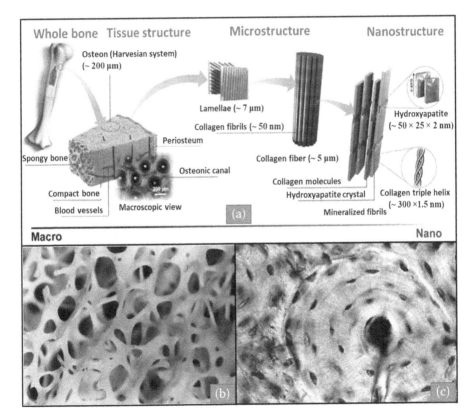

**FIGURE 9.1** Pictures of Different Parts of the Bone (a) (Sadat-Shojai et al. 2013), (b) Microstructure of Cancellous Bone, (c) Microstructure of Cortical Bone (Reprinted from *Journal of Biomechanics*, 32 (11), Fyhrie, David P and JH Kimura, Cancellous Bone Biomechanics, 1139–1148, 1999, with kind permission from Elsevier).

periosteum, the compact and dense part of the bone is known as compact or cortical bone, and the inner part that is highly porous (porosity 70%–90%) and spongy in nature is known as cancellous bone. Cancellous bone has an interconnected 3-D network of the pores in which tissues grow and the transportation of body fluids takes place. A highly porous structure houses the bone marrow, blood vessels and other organic nutrients. The formation of red blood cells, certain white blood cells and platelets (hematopoiesis) also happen in the soft core of the bone marrow (White, Black, and Folkens 2011). Cortical bone has a very dense structure (with a porosity of 5%–40%) and acts as a protective wall for the inner cancellous bone. It facilitates the transportation of body fluid through Haversian and Volkmann's canal. Haversian canals are the series of micro-tubes surrounded by the concentric lamellae jointly known as osteons (Rho, Kuhn-Spearing, and Zioupos 1998). In the macroscopic view, it can be visualised that the diameter of the osteon is approximately 200 μm. Volkmann's canals are the tiny interconnected channels that transport the blood, and these are interconnected to the Haversian canals. At the micro-level, bones are made of plate-shaped lamellae which are a combination of cylindrical collagen fibres. Collagen is a type of insoluble protein which is abundantly found in the animal kingdom. The molecules of collagen are packed together in the form of plate-shaped hydroxyapatite crystals. These crystals are assembled in layers that form thin, cylindrical fibrils. Collagen fibres are the conjoint form of these cylindrical fibrils.

The microstructures of cancellous and cortical bone are shown in Figure 9.1(b) and (c) respectively (Fyhrie and Kimura 1999). For cancellous bone, the volume fraction of hard tissue is very low as compared to cortical bone. The shape of the hard tissues is roughly cylindrical and are attached to one another to form

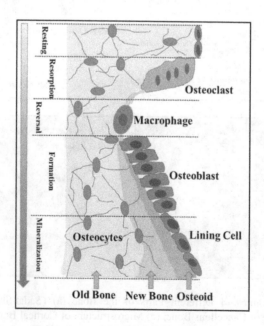

**FIGURE 9.2**   Restoration Process of New Tissues of the Bone.

interconnected cavities. These cavities host the bone marrow and provide passage for the circulation of biofluid. For cortical bone, the volume fraction of hard tissues is higher as compared to the cancellous bone. Haversian canals in the cortical bone are also long and close to cylindrical in shape. Blood vessels and nerves travel through the Haversian canal.

## 9.2.2 Bone Regeneration Process

There are three main types of cells that are found in the bone, as shown in Figure 9.2. The first is the osteoblast which is responsible for the formation of bone. Bone resorption or destruction is caused due to the osteoclast and osteocytes cells, which are basically osteoblast restrained in the matrix of the bone for maintenance. Bone cells react according to the mechanical forces in the form of strain (Frost 1998). It is assumed that osteocytes work as the mechano-sensor that detects mechanical signals. At resting position (zero strain), the generation of osteoclast takes place, and this reduces bone density and resorbs the bone. At proper strain experienced by the osteocytes, osteoblast cells generate and strengthen the bone (Sims and Gooi 2008). The functioning of different cells of the bone and the restoration of new bone tissues are shown in Figure 9.2.

## 9.2.3 Mechanical Properties of the Bone

Human bone is an anisotropic material; therefore, its properties vary with the direction of loading. Compressive stress-strain curves for the cancellous and cortical bone are shown in Figure 9.3(a) and (b) (Shim et al. 2005; Wang et al. 2020; Öhman et al. 2011). For the compression test, specimens of cancellous bone are acquired from the cervical spines of eight male corpses of different ages ranging from 40 to 79 years. Meanwhile, cortical bone specimens were procured from femoral or tibial shafts. Cortical bone specimens were divided into two groups (I - child group: age range 4 to 15 years and II – adult group: age range 22 to 61 years) in order to investigate the effect of age on compressive strength.

The ultimate compressive stress (i.e. Point C in Figure 9.3(a)) maximum stress point is attained by the material in a compression test. The stress point at which pore of the bone starts to become permanently damaged is known as yield stress (Point B in Figure 9.3(a)). This yield stress is also defined as the stress from which the stress-strain curve becomes non-linear. The stress-strain diagram of the bone can be divided into two distinct regions, known as the elastic region (AB) and the plastic region (BCD). These two regions are divided by the point of yield strength. In the elastic region, the pores of the bone are elongated elastically without any permanent deformation, and it retains its original dimension and shape after unloading. The region in which stress does not linearly increase and oscillates along definite stress is referred to as the plastic region. In this region, pores of the bone collapse per-manently, and bone does not regain its shape and size. The point at which bone fractures is known as the point of rupture, and the stress at this point is termed the ultimate strength of the bone.

The trend of the stress-strain curve is similar for both cancellous as well as cortical bone. The yield strength and ultimate strength for the cortical bone is comparatively higher than in cancellous bone. However, the yield elongation and the ultimate elongation point for cancellous bone are higher than the cortical bone. This is caused by the presence of fine lamellae in the cancellous bone that can enhance elongation but reduce strength due to lower load-bearing capacity. Below the yield point, bone is safe. Therefore, the yield strength is considered as the strength of the bone.

The compressive strength of the bone also depends on the age of a person. As observed in Figure 9.3(b), the density of adult cortical bone ($\rho_{ad}$) is about 15%

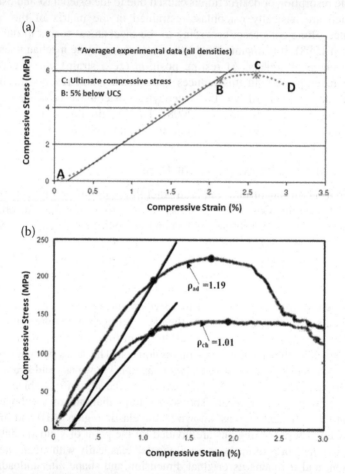

**FIGURE 9.3** A Characteristic Stress-Strain Diagram of the Bone Material. (a) Cancellous Bone (Reprinted from the *International Journal of Impact Engineering*, 32 (1–4), Shim, VPW, LM Yang, JF Liu and VS Lee, Characterisation of the Dynamic Compressive Mechanical Properties of Cancellous Bone from the Human Cervical Spine, 525–540, 2005, with kind permission from Elsevier) (b) Cortical Bone (Reprinted from Bone, 49 (4), Öhman, Caroline, Massimiliano Baleani, Carla Pani, et al., Compressive Behaviour of Child and Adult Cortical Bone, 769–776, 2011, with kind permission from Elsevier).

**TABLE 9.2**
**Yield Compressive Strength and Yield Tensile Strength with Elastic Modulus for the Cortical and Cancellous Bones**

| Bone Type | Elastic Modulus (GPa) | Yield Strength in Compression (MPa) | Yield Strength in Tension (MPa) | References |
|---|---|---|---|---|
| cortical bone | 14.1–27.6 | 219 ± 26 longitudinal 153 ± 20 transverse | 172 ± 22 longitudinal 52 ± 8 transverse | (Snyder and Schneider 1991; Guo 2001) |
| cancellous bone | 0.1–0.4 | 1.5–9.3 | 1.6–2.42 | (Eckardt and Hein 2001; Ford and Keaveny 1996; Rho, Ashman, and Turner 1993) |

higher than the density of a child's cortical bone. Moreover, the yield strength and elastic modulus of children's cortical bone are 38% and 34% lower than the adult cortical bone.

The yield strength with their corresponding elastic modulus at different loading conditions (longitudinal and transverse) for both the cortical and cancellous bone at compressive and tensile loading state is listed in Table 9.2. From this table, it can be concluded that both types of bone are weak in tension.

## 9.3   BONE SCAFFOLD, GRAFTING OR IMPLANTATION

In case of severe damage to the bone tissues due to the injury caused by an accident or a bone degenerative disease, natural regeneration of the bone tissues might not heal the affected bone in an adequate manner. Hence, grafting is required for sufficient healing. Different types of grafting which are in current practice include the following:

### 9.3.1   AUTOGRAFTING

In the autografting process, injured bone tissues are replaced by the extraction of healthy bone tissues from the body of the patient. This is the ideal method in repairing the affected tissues because of its high compatibility and lower chances of rejection (Leukers et al. 2005). However, due to the removal of bone tissues, additional surgery is required for the patient, and the removed parts of the body will always be vacant. Therefore, this is useful only for small injuries or bone imperfection.

### 9.3.2   ALLOGRAFTING

Allografting is the process of inserting a collection of processed bone tissues into the patient's injured bone. However, there are certain additional challenges faced by the recipient, such as the possible lack of osteopotency preservation during the storage of the material. Moreover, inherent diseases can be transmitted from donor to recipient.

### 9.3.3 Bone Grafting through Tissue Engineering

The treatment of sick or injured bone can also be done by grafting with biomaterials that are designed to adapt to the environment of natural bone tissues. The design of artificial material for the bone is referred to as tissue engineering, while the designed material is known as the bone implant or scaffold. Desirable three-dimensional material should be biocompatible and must not harm the host bone tissues when implanted. The material is formed in such a way that it can mimic the natural bone in the form of strength, porosity and the topography of the pores (Kim et al. 2017; Butscher et al. 2011). Biodegradable and stable or non-degradable materials are generally used for manufacturing bone implants. Biodegradable materials work as support for the generation of tissues. These bone tissues are anchored on the surface of the biodegradable implant. Simultaneously, the implant degrades while in body fluid. Mature bone tissues take the place of the implant after its degradation. Another type of implant is made by a stable material known as the stable implant. It is not biodegradable and remains to persist in the body after implantation.

## 9.4 MATERIALS FOR BONE IMPLANT

### 9.4.1 Biodegradable Materials

#### 9.4.1.1 Polymeric Scaffold

For the fabrication of bone implants, polymers are generally used due to their considerable flexibility in design. These can be tailored to any shape, and their mechanical properties can be matched with the necessary requirement (Nandakumar et al. 2010). The biodegradability of polymers depends on molecular design, hydrophobic and hydrophilic behaviour in the aqueous environment of body fluid and content of enzymes and cellulose (Tang et al. 2016; Huang et al. 2019). Therefore, polymers are universally studied in the application of tissue engineering. Polymers can be categorised in two: natural polymers and synthetic polymers.

Due to the exceptional biocompatibility and bio-functionality of molecules present on the surface of natural polymers, these are of great relevance to tissue generation research. Biomolecules help in the integration of newborn tissues to their surface (Saranya et al. 2011). Silk, alginate, chitosan and collagen have been investigated for the use of bone scaffold. Some investigations claim that natural polymers extracted from the bovine Achilles tendon show higher cell viability without any cytotoxicity (Nocera et al. 2018; Li, Wu, et al. 2019). However, the risk of disease transmission from natural polymers limits its suitability. In addition, insufficient mechanical properties also inhibit appropriate bone scaffold.

To overcome the insufficient mechanical properties of natural polymers, natural polymer composites were fabricated, in which hydraulic acid-based, chitosan-based, collagen-based composites are important. Fibres of acrylonitrile butadiene styrene (ABS) were added to collagen in order to produce the scaffold using the 3D printing method (Mozdzen et al. 2016). It was discovered that the elastic modulus of the scaffold increases up to 68-fold in comparison to the collagen scaffold. In another observation, pure chitosan scaffold was compared with chitosan-based composite reinforced with cross-linked pectin and genipin, It was identified that the strength of the composite greatly increases with lower degradability that can also enhance the proliferation rate of osteoblast (Liu, Chang, and Lin 2015).

Nowadays, synthetic polymer scaffold is a subject of interest due to their customisable properties like porosity, strength and elastic modulus (Ge et al. 2009). The prime advantage of the polymeric scaffold is that it can be manufactured in any shape and size by the use of 3D printing technology. Also, the toxicity of the degraded product is categorically low, and this is useful to metabolism. Polylactic acid (PLA), polyglycolic acid (PGA), polycaprolactone (PCL) and polyurethane (PU) are the most common materials that are used for bone implant application (Sun et al. 2017; Lasprilla et al. 2012; Angeloni et al. 2017; Williams et al. 2005).

Synthetic polymeric materials like PLA, PGA, PCL and PU can be also used as matrix materials to synthesise the polymer composite by adding different reinforcing agents in order to enhance the strength of the polymeric materials used to obtain improved bone regenerative properties. For the pure PCL bone implant, mechanical properties are inadequate; therefore, it was reinforced with hydroxyapatite (HA) and $\alpha$-Tricalcium phosphate (TCP) (Vella et al. 2018). It was discovered that flexural strength and modulus of this PCL scaffold increase due to the addition of these materials, but fracture toughness was almost unchanged. In another study, it was determined that the addition of calcium phosphate as a reinforcement in the PCL considerably improves the compressive strength of the bone scaffold, and this also showed cytocompatibility with newborn bone tissues (Mondrinos et al. 2006). To make the PLA composite, it was coated with mussel-inspired polydopamine (PDA) to be able to investigate the biological response of this material. It was observed that the PLA/PDA composite enhances the adhesion property of the cells and heightens the proliferation rate (Kao et al. 2015).

### 9.4.1.2 Metallic Scaffold

The polymeric scaffold has adequate biocompatibility and plasticity. However, poor hydrophilic characteristics, lower compressive strength as compared with natural bone and risk of aseptic inflammation prevent it from being the ideal bone scaffold (Han, Liu, et al. 2012; Shuai et al. 2019; Wegener et al. 2011). To obtain the desired strength with enhanced biological performance, some metallic scaffolds are developed in which Mg and Mg alloys, Zn and its alloys and Fe and its alloys are proposed for the application of bone implants.

#### 9.4.1.2.1 Mg-Based Alloys

Mg and its alloys are useful materials for bone implants because of the similarity in physical and mechanical properties of the Mg alloy with the natural bone (Crespi,

Capparè, and Gherlone 2009). Mg is an essential element in the body that helps in maintaining the function of nerves and muscles to make the bones strong. The young modulus of Mg varies between 10 to 30 GPa, and that is nearly equal to the young modulus of cortical bone (Kirkland and Birbilis 2014; Witte et al. 2005). However, the corrosion resistance of Mg is prominent in the medium of body fluid that makes it highly degradable (Staiger et al. 2006). Therefore, its mechanical properties reduce significantly fast and the premature failure of the Mg implant takes place before adequate tissue healing is allowed to process (Chakraborty Banerjee et al. 2019). It is a major drawback of the Mg implant. For proper in-growth of the newborn bone tissues, the degradable rate of the implant should be slow. The corrosion resistance of Mg can be regulated by alloying it with suitable elements. Different Mg alloys have been studied to improve the biological, mechanical as well as corrosion properties of implants. The alloying of Mg with Al and Ca in appropriate proportions enhances the creep strength as well as the yield strength of the material due to the formation of hard interconnected Laves phase (Zubair et al. 2019). In another study, it was determined that the addition of Al in Mg reduces the corrosion resistance of the alloy due to the formation of micro-galvanic cells (Xin et al. 2007). Corrosion resistance, as well as fatigue life of the Mg alloy (AZ31), can be also be improved by refining the grain through the hot rolling process (Wang et al. 2007). To increase the corrosion resistance and bio-compatibility of Mg, it is coated with hydroxyapatite (HA). However, due to the brittle quality of HA, the material that reduces the corrosion resistance cracks. Furthermore, this Mg-HA composite is coated with hybrid polyetherimide (PEI)-$SiO_2$. It was found that due to coating with PEI-$SIO_2$, corrosion resistance and bioactivity of the material considerably developed (Kang et al. 2019). However, the ideal degradability rate could not be achieved until the present times, when the best tissue ingrowth results and experiments are ongoing in order to achieve the desired properties.

### 9.4.1.2.2 Fe-Based Alloys

The mechanical properties of Fe alloys surpass those of the Mg alloys' and are suitable for bone implant application. Unlike an Mg alloy, the degradation rate of Fe-based alloy is too slow. Therefore, there is current research on how to opti-mise the degradation rate of the Fe alloy to be able to fully compensate the requirement of the generation of matured bone tissues. Initially, Mn is added as the alloying element in order to accelerate the degradation rate and reduce the ferromagnetic property of the Fe (Hermawan et al. 2008). This also improves the strength of the material. The addition of Mg in Fe also improves the degradability rate of the alloy due to the presence of a more electrochemically active element. It forms the micro-galvanic cell with Fe and reduces the potential of the alloy (Oriňáková et al. 2013). Cheng and Zheng synthesised the Fe-W alloy by adding different wt % (i.e. 2% and 5%) of W in the Fe matrix (Cheng and Zheng 2013). The addition of W reduces the corrosion resistance of the alloy. In some other studies, Ag and Au were used as alloying elements in the Fe matrix, and it was discovered that the degradation of Fe increases due to the formation of galvanic cells in which Fe works as an anode (Huang et al. 2016). Although the corrosion rate of Fe can be manipulated by the alloying with appropriate elements, Fe particles accumulate in the various sites of tissues that cause inflammation in the body. Therefore, it is essential to seek a suitable material that can be satisfactorily employed as bone implants.

### 9.4.1.2.3 Zn-Based Alloy

The degradation rate of the Zn-based alloy is lower than the Mg-based alloy but higher than the Fe-based alloy. Therefore, it represents a moderate degradation rate that may be useful in the application of bone implants. The main consideration of Zn is poor mechanical properties, and the tensile strength of as-cast Zn is near 30 MPa (Li, Zheng, and Qin 2014). Also, the ductility of the Zn is exceedingly poor due to its HCP structure. These entities do not meet the requirements for use as a bone scaffold. Because of the high solubility of Mg in the Zn, their alloy is prepared to enhance mechanical properties. Kubasek et al. reported that by the addition of Mg in different wt % (0% to 1.6%), the ultimate strength of Zn increases significantly from 120 MPa to 301 MPa with improved hardness (Kubásek et al. 2016).

## 9.4.2 NON-DEGRADABLE MATERIALS

Non-degradable materials as bone scaffolds are permanent and insoluble in the bodily fluid. The degradability rate of these types of materials is profoundly slow in the presence of the body fluid medium. Certain metallic materials and ceramics are non-degradable and are widely used for bone implants.

### 9.4.2.1 Ceramic Materials

The compositions of hydroxyapatite (HA) and CaP (calcium Phosphate) are nearly similar to natural bones (Eliaz and Metoki 2017). These are bioactive ceramics, and chemical bonding takes place between these materials and the neighbouring bone. Moreover, they have good biocompatibility, osteoconductivity and osteointegration to the newborn bone tissues which facilitate the enhancement of healing rate. However, due to high crystallinity and good stability, the degradability rate of these bioactive ceramic materials is very slow. Hence, a subsequent operation is required in order to remove the implant after the injury heals.

### 9.4.2.2 Stainless Steel

Stainless steel is an iron-based alloy containing 11% to 30% chromium as the main alloying element, along with trace amounts of nickel (Jung et al. 2018; Zhang et al. 2019). It is good in terms of biocompatibility and has high corrosion resistance. The presence of Cr in stainless steel provides superior passivation on its surface and enhances corrosion resistance. Furthermore, austenitic steel is nonmagnetic in nature and does not provide any artefacts in magnetic resonance imaging (MRI). However, 316L stainless steel is attacked by pitting corrosion and stress corrosion cracking that causes the release of alloying elements such as Ni, Cr and Mo (Kuroda et al. 2002). Ni is a carcinogenic element that can bring forth health problems upon its release. Moreover, the young modulus of an elastic modulus is significantly higher than the natural bone, and this causes the stress shielding effect; therefore, bone resorption takes place. Due to high elastic modulus and a smooth surface, anchoring between the newborn tissues and the implant does not occur, and this causes the loosening of the implant and replacement. Therefore, it is necessary to change this material into one that is more compatible with natural bone.

### 9.4.2.3   Co-Cr Alloy

Co-Cr alloy is generally used in joint replacement, such as femoral head, knee replacement and hip replacement. It is employed as the screw used for combining the implant with the natural bone due to enhanced hardness and good wear resistance (Hussain, Sah, and Sidebottom 2014). Also, osteointegration of Co-Cr alloy is lower; therefore, it is easy for removal after the injury heals. The strength of the Co-Cr alloy is higher than the durability of stainless steel, but this alloy is significantly affected by galvanic corrosion due to the presence of two dissimilar metals in higher quantities. Cr can provide the passivation layer on the surface of the alloy, but the removal of this layer due to fretting action on the joint can accelerate corrosion. Finally, poor osteointegration with the neighbouring natural bone and the neutrality of the attachment with newborn tissues prevent it from becoming an acceptable bone implant material (Bahrami, Fathi, and Ahmadian 2015).

### 9.4.2.4   Ni-Ti Alloy

Ni-Ti alloys demonstrate shape memory effects, superelasticity with high mechanical properties and an elastic modulus close to natural bone (Dudek and Goryczka 2016). Therefore, it can be used in bio-implant applications. Although there is a presence of Ti as a naturally occurring oxide layer on the layer of alloy, it is not enough to prevent the effects of biological fluid. Hence, the debris of Ni particle falls into the biological fluid and contaminates it. The presence of Ni ions in the bio-fluid causes health issues such as hypersensitivity, pneumonia, chronic sinusitis and lung cancer (Michalska et al. 2017). Therefore, it is necessary to replace this alloy with a suitable material in order to overcome these problems.

### 9.4.2.5   Titanium and Its Alloys

Titanium and its alloys can be used as biomaterials due to their distinctive properties. CP Ti is less cytotoxic and carcinogenic; hence, it is biocompatible with the human body (Gepreel and Niinomi 2013). An experiment was performed in order to investigate the cytotoxicity of the metals used for orthopaedic application. It was observed that the activity of macro-phases (i.e. large cells that fight against infection) of murine are unaffected by pure titanium implants (Rae 1975). It has been also shown that there is no toxic effect on the fibroblast of the patient using titanium as an implant (Rae 1981). The corrosion resistance of titanium implant in the bodily fluid is higher than the other known biocompatible metals due to the formation of a thin layer of oxide on the surface of the implant that prevents the dissolution characteristic of the titanium atoms in the medium of body fluid. Numerous studies were performed on the electrochemical property in order to investigate the behaviour of the titanium implant, and this proved outstanding corrosion resistance (Speck and Fraker 1980; Brune, Evje, and Melsom 1982). Tissue ingrowth is enhanced on the rough surface of the implant. Koike and Fujii observed that the surface roughness of the titanium alloy implants improves the growth of tissues (Koike and Fujii 2001). The advantageous properties of titanium alloys such as good biocompatibility, tremendous corrosion resistance, high specific strength and roughness of the surface make it suitable for use as bone implants in the human body.

In the present time, CP Ti and α+β alloys are commonly used as structural as well as functional biomaterials for the replacement of hard bone tissue, such as knee and hip joints, femoral bones and dental implants. These materials have excellent specific strength with high corrosion resistance without producing any artefact in the body. The alloys that are usually used in the bio implant application are Ti-3Al-2.5V, Ti-5Al-2.5Fe, Ti-5Al-2Mo-2Fe, Ti-5Al-3Mo-4Zr, Ti-13Nb-13-Zr, Ti-29Nb-13Ta-3.6Zr and Ti-45Nb. However, Ti alloys can be used in the application of bone implantation but there persist some major problems such as stress shielding effects due to the mismatch of the elastic modulus of natural bone and implants causing bone resorption (Elmay et al. 2013). Stress shielding is the process of bone density reduction due to improper mechanical signals acquired by the osteocyte bone tissues. An external load is carried by the bone. When a prosthesis or an implant with a higher elastic modulus than the natural bone is introduced to the body, the external load is shared by the implant as well as the natural bone. Therefore, reduced stress is experienced by the bone as compared to normal conditions, causing stress shielding (Huiskes, Weinans, and Van Rietbergen 1992).

## 9.5 INTRODUCTION OF POROUS IMPLANT AND SYNTHESIS PROCESS

The elastic modulus of the natural bone varies between 0.05 to 30 GPa, while it varies between 90 to 200 GPa for metallic biomaterials scaffolds such as Ti alloys and stainless steel (Long and Rack 1998; Han, Meng, et al. 2012). For proper working of the bone implant, its elastic modulus should be matched with the natural bone to avoid the stress shielding effect. Furthermore, the implant should be strong enough to bear the stresses present as a bone implant. The elastic modulus of the implant can be paired with natural bone by the incorporation of pores in the bulk material. According to the Gibson-Ashby model, elastic modulus directly depends on the density of the porous material (Ashby et al. 2000).

$$\frac{E_f}{E_s} = \alpha \left( \frac{\rho_f}{\rho_s} \right)^n \tag{9.1}$$

Where, $E_f$ = elastic modulus of the porous material, $E_s$ = elastic modulus of the dense material, $\rho_f$ = density of the porous material, $\rho_s$ = density of the dense material and $\alpha$, n = constant and fixed exponent respectively.

Therefore, according to Equation (9.1), the elastic modulus can be specially made by controlling the density of the porous material. With the reduction of density, elastic modulus, as well as the yield strength of the product, reduces (Singh, Abhash, Yadav, et al. 2019).

The porous structure of the biomaterials also promotes bone ingrowth and can increase the healing rate of the injury (Yang et al. 2001). Besides this, pores are the favourable anchorage sites for the newborn tissues which reduce the chance of implant loosening (Abbasi et al. 2020). In the last decades, different fabrication techniques were developed in synthesising porous bone implants. Generally, loose powder sintering, freeze casting, the space holder technique and additive manufacturing are adopted to produce the porous materials (Singh, Singh, and Mondal 2019; Abhash, Singh, Kumar, et al. 2020; Torres et al. 2014; Jenei et al. 2016; Parthasarathy, Starly, and Raman 2011). The individual descriptions of the synthesis processes are as follows:

### 9.5.1 LOOSE POWDER SINTERING

Loose powder or gravity sintering is generally used for bronze, stainless steel, titanium and superalloys. A schematic diagram of loosely packed metal powder sintering is shown in Figure 9.4. The powders of the metals are loosely packed in a chamber and allowed to sinter. During sintering, atoms diffuse among the particles of the powder, and bonding takes place. Generally, the loose bronze powder is sintered at 820°C and 20% to 50% porosity levels are achieved, but the strength of the sintered product is lower. The strength can be increased by compaction or rolling the packed powder prior to sintering, but the porosity of the product is reduced (Nouri, Hodgson, and Wen 2010).

### 9.5.2 GEL CASTING

Gel casting was initially used for processing the porous ceramic product of high porosity and interconnected pores (Gilissen et al. 2000). The procedure of foam synthesis by this method is shown in Figure 9.5. In this process, the slurry is prepared by mixing the metal powder, a dispersing agent and an organic monomer in a solvent. Generally, distilled water is used as the solvent (Sepulveda et al. 2000). The porosity of the foam depends on the process parameters such that the amount of organic monomer, dispersing agent, initiator, etc. The mixture is then properly homogenised. A mechanical stirrer is used to make the mixture uniform. Ball milling can be also used to obtain high evenness. The polymerisation process is started by adding the appropriate initiator in the homogenised mixture (certain catalysts are also added to the mixture prior to the addition of the initiator) and further stirring is performed. In situ, the polymerisation process takes place because of the cross-linking reaction in the mixture due to the presence of an initiator in order to produce the foamed gel. After thoroughly stirring, the mixture is then poured into a mould that is used for casting in a specific shape and allows cooling at room temperature. The cast product is removed from the mould and dried at a higher temperature for the evaporation of the solvent. After drying, the green product is capable of binding with the powder to retain the shape of the cast due to the polymer as formed by polymerisation. Green products are further sintered in the appropriate furnace at a suitable temperature and time.

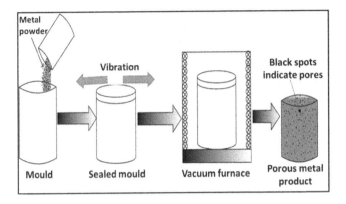

**FIGURE 9.4** Schematic Diagram of Loose Powder Sintering Method.

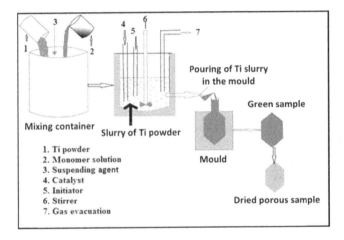

**FIGURE 9.5** The Process of Gel Casting for the Processing of Porous Titanium Alloy.

## 9.5.3 SPACE HOLDER TECHNIQUE

The space holder technique is a popular method used to produce pores in materials. A space holder is a fugitive material in the form of particulates that can be evaporated at moderate temperature or leached out in an appropriate solvent. Ammonium bicarbonate (Abhash, Singh, Muchhala, et al. 2020), urea (Sharma et al. 2011) and magnesium (Aydoğmuş and Bor 2009) are evaporative types of space holders. These materials can be evaporated at a temperature below 600°C. Sodium chloride (Jha et al. 2013), saccharose (Jakubowicz, Adamek, and Dewidar 2013) and carbamide powder (Arifvianto, Leeflang, and Zhou 2014) are dissolvable or leach-out types of space holders and can be easily drained out in warm water.

Volume fraction, size and shape of the pores in the porous product can easily be manipulated with the selection of the appropriate amount, shape and size of the space holder particulates (Abhash, Singh, Ch, et al. 2020; Singh, Abhash, Nair, et al. 2019). A schematic diagram used to produce a porous metal product is illustrated in Figure 9.6. First, an appropriate amount of space holder particulates is properly mixed with metal powder in an automatic tumbler. After mixing, the metal powder and space holder mixture are transferred into a hardened steel die and compacted using a hydraulic press. The compacted mixture part is removed from the die after the completion of compaction. The compressed part is known as the green compact. If the used space holder is an evaporative type, then the green compact is dried in a vacuum oven at an appropriate temperature in order to evaporate the space holder. Therefore, after drying, the pores are created in the green compact in place of the space holder particulates. Finally, the dried compact is sintered in a high-temperature vacuum furnace to obtain strength in the sintered product.

If the used space holder is the dissolvable or leach out type, then the drying of the green compact is avoided is instead directly sintered in the furnace for an appropriate time and temperature. After sintering, the sintered product is immersed in an appropriate solvent to leach out the space holder particulates, and therefore, spaces or pores are created in the product.

Figure 9.7 represents the microstructures of porous Ti alloys that are synthesised through the space holder technique using varying volume fraction ammonium

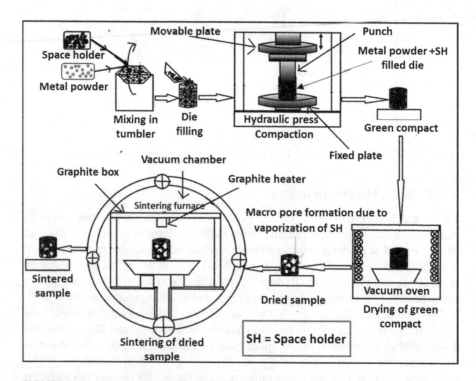

**FIGURE 9.6** Schematic Diagram for the Synthesis of Porous Product through the Space Holder Technique.

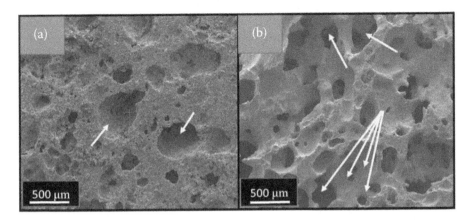

**FIGURE 9.7** Microstructures of a Porous Ti Alloy. (a) is the microstructure for the porous product made by using 50 v% of the space holder, while (b) represents the microstructure for the porous sample synthesised using 80 v% of the space holder. (Reprinted from *Materials Science and Engineering: C*, 98, Singh, Pradeep, IB Singh and DP Mondal. A Comparative Study on Compressive Deformation and Corrosion Behaviour of Heat Treated Ti4 wt % Al Foam of Different Porosity Made of Milled and Unmilled powders, 918–929, 2019, with kind permission from Elsevier).

bicarbonate as a space holder (Singh, Singh, and Mondal 2019). Figure 9.7(a) shows the microstructure of porous Ti alloy made by the 50 v% of ammonium bicarbonate, while Figure 9.7(b) displays the microstructure for the product prepared by the use of 80 v% of the ammonium bicarbonate. From these figures, it can easily be observed that the number of pores is higher in the product made by using 80 v% of ammonium bicarbonate (a single arrow represents the pores). Moreover, interconnectivity among the pores increases with each increment of the applied space holder amount in order to make the porous product (interconnected arrows show the interconnectivity).

### 9.5.4 POROUS PRODUCT SYNTHESISED THROUGH ADDITIVE MANUFACTURING

Nowadays, the prominence of additive manufacturing (AM), which is also known as three-dimensional (3D) printing and rapid prototyping, is rising in the applications of tissue engineering, specifically in the production of unique and patient-specific biomedical implants. AM has a great advantage over conventional manufacturing processes in terms of easy product manufacturing, controlled geometry and pore distribution and increased productivity and reduced labour cost (Yang et al. 2002). It operates through the combination of computer-aided design (CAD) and computer-aided manufacturing (CAM) (Hutmacher, Sittinger, and Risbud 2004). Since the discovery of AM more than 30 years ago, a number of modifications have been made until the present in order to tackle the complicated design and specific requirements of the product. Powder bed fusion (PBF), fused deposition modelling (FDM), stereolithography (SL) and direct energy deposition (DED) are the main methods of AM. The composition and microstructural control in the product manufactured through DED technique is excellent. Therefore, porosity and cracks are annihilated; hence, improved mechanical properties are possible. However, some of the drawbacks of this technique include insignificant dimensional tolerance and poor surface finish.

The PBF technique is used to manufacture a 3D product with excellent mechanical properties and good dimensional accuracy. A bone implant with controlled porosity and pore size can easily be synthesised through this technique. To achieve this, a 3D CAD model is created by the selection of the appropriate porous unit cell. Cubic (Kadkhodapour et al. 2015), diamond (Li, Jahr, et al. 2019), gyroid (Yan et al. 2015), octahedral (Yang et al. 2019), etc. unit cells are used to produce the porous 3D CAD model. In the PBF process, metal powder is used as a raw material to manufacture the product. A thin layer of metal powder is spread on the pre-sintered layer with the help of a roller or blade. Furthermore, this layer is fused by the use of a laser or an electron beam. Selective laser sintering (SLS) and selective laser melting (SLM) are the two techniques adopted to fuse the material. In the SLS process, materials are energised by the laser beam below the melting point, and diffusion happens among the powder particles. Meanwhile, in the SLM process, materials in powder form are fully melted to consolidate in the form of the final product. Metallic biomaterials such as Ti alloy, Co-Cr alloy, steel, etc. are consolidated through the SLM process (Lee et al. 2017). A schematic diagram of SLM is shown in Figure 9.8. At the beginning of the process, a thin layer of the metal powder is spread on the building platform with the use of a roller, and a laser beam with sufficient energy heats the powder in the selected area. Due to heating, the powder melts, and a liquid pool forms. This quickly cools down, and the manufacturing of the product starts. After forming a layer, the building platform moves downward by a height that is equal to the thickness of the layer. Simultaneously, the piston delivers the metal powder, which is again deposited on the building platform by the roller, and the process is repeated until the completion of the product.

## 9.6  SUMMARY

The development of biomaterials with properties analogous to natural bone is the subject of investigation, and extensive research is happening in the present times. In the early stage of implantation, natural polymers like alginate, chitosan, collagen etc. are used due to better attachment with fibrous tissues, shape flexibility and biodegradability. However, the lack of suitable mechanical properties and susceptibility to the transmission of inherent disease from donor to recipient prevent it from becoming an ideal biomaterial. Certain synthetic polymers such as PLA, PCL and PU were also tested as bone materials, but these have inferior mechanical properties. Some metallic materials like Mg, Fe, Zn and their alloys that have adequate biomechanical properties and are degradable in the presence of biofluid were also evaluated as bone implant materials.

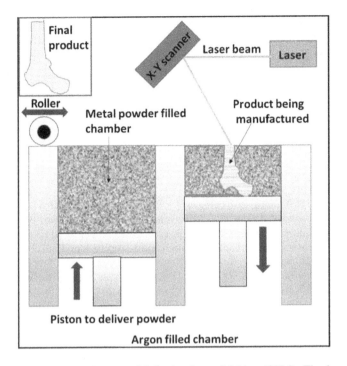

**FIGURE 9.8**   Schematic Diagram of Selective Laser Melting (SLM). The layer-by-layer formation of the product happens due to the melting of the metal powder in the selected area. The powder is melted because the laser beam energy falls over the metal powder.

However, their degradation rate does not match with the healing rate of bone injury. Hydroxyapatite and calcium phosphate are ceramic materials that have a composition that is similar to bone, and these also have good bioactive properties with the bone tissues. On the other hand, their poor ductility, brittleness and non-degradable nature are undesirable. In the present time, non-degradable Ti and its specific alloys are the subjects of attention as the optimal bone implant material due to its good biocompatibility, light weight and higher corrosion resistance. However, the higher elastic modulus of Ti alloys as compared to natural bone is a major reason for implant failure, specifically due to the stress shielding effect. It can be synchronised with natural bone with the incorporation of pores in the metallic implant. Various processes such as the space holder technique, loose powder sintering, gel casting, additive manufacturing etc. are discovered to produce pore in the bulk material and for the matching of elastic modulus with bone. Loose powder sintering and gel casting provide limited porosity. About 90% porosity can be created in the product through the space holder technique, but controlling the size and shape of the final product is difficult. Additive manufacturing is the advanced technique used to synthesise the highly porous product with excellent accuracy in pore shape and size. Also, the additively manufactured porous product has superior dimensional accuracy than other processes. Though a number of biomaterials that can partially mimic the natural bone are developed, the ideal bone implant has not been discovered until now. At present, research on the invention of the excellent bio implant which can perfectly stimulate the bone tissues and reduce the healing time of the injury is still ongoing.

# REFERENCES

Abbasi, Naghmeh, Stephen Hamlet, Robert M. Love, and Nam-Trung Nguyen. 2020. Porous scaffolds for bone regeneration. *Journal of Science: Advanced Materials and Devices* 5: 1–9.

Abhash, Amit, Pradeep Singh, Venkat A.N. Ch, Sriram Sathaiah, Rajeev Kumar, Gaurav K. Gupta, D.P. Mondal. 2020. Study of newly developed Ti–Al–Co alloys foams for bioimplant application. *Materials Science and Engineering: A* 774:138910.

Abhash, Amit, Pradeep Singh, Rajeev Kumar, et al. 2020. Effect of Al addition and space holder content on microstructure and mechanical properties of $Ti_2Co$ alloys foams for bone scaffold application. *Materials Science and Engineering: C*110600.

Abhash, Amit, Pradeep Singh, Dilip Muchhala, Rajeev Kumar, Gaurav K. Gupta, and D. P. Mondal. 2020. Research into the change of macrostructure, microstructure and compressive deformation response of $Ti_6Al_2Co$ foam with sintering temperatures and space holder contents. *Materials Letters* 261: 126997.

Angeloni, Valentina, Nicola Contessi, Cinzia De Marco, et al. 2017. Polyurethane foam scaffold as in vitro model for breast cancer bone metastasis. *Acta Biomaterialia* 63: 306–316.

Arifvianto, B., M. A. Leeflang, and J. Zhou. 2014. A new technique for the characterization of the water leaching behavior of space holding particles in the preparation of biomedical titanium scaffolds. *Materials Letters* 120: 204–207.

Ashby, Michael F., Tony Evans, Norman A. Fleck, J. W. Hutchinson, H. N. G. Wadley, and L. J. Gibson. 2000. *Metal foams: a design guide*. Oxford, UK: Elsevier.

Aydoğmuş, Tarık, and Şakir Bor. 2009. Processing of porous TiNi alloys using magnesium as space holder. *Journal of Alloys and Compounds* 478 (1–2): 705–710.

Bahrami, M., M. H. Fathi, and M. Ahmadian. 2015. The effect of nanobioceramic reinforcement on mechanical and biological properties of Co-base alloy/hydroxyapatite nanocomposite. *Materials Science and Engineering: C* 48: 572–578.

Banerjee, Parama Chakraborty, Saad Al-Saadi, Lokesh Choudhary, Shervin Eslami Harandi, and Raman Singh. 2019. Magnesium implants: Prospects and challenges. *Materials* 12 (1): 136.

Brune, Dag, Dag Evje, and Sigurd Melsom. 1982. Corrosion of gold alloys and titanium in artificial saliva. *European Journal of Oral Sciences* 90 (2): 168–171.

Butscher, A., M. Bohner, S. Hofmann, L. Gauckler, and Ralph Müller. 2011. Structural and material approaches to bone tissue engineering in powder-based three-dimensional printing. *Acta Biomaterialia* 7 (3): 907–920.

Cheng, J., and Y. F. Zheng. 2013. In vitro study on newly designed biodegradable Fe-X composites (X = W, CNT) prepared by spark plasma sintering. *Journal of Biomedical Materials Research Part B: Applied Biomaterials* 101 (4): 485–497.

Crespi, Roberto, Paolo Capparè, and Enrico Gherlone. 2009. Magnesium-enriched hydroxyapatite compared to calcium sulfate in the healing of human extraction sockets: Radiographic and histomorphometric evaluation at 3 months. *Journal of Periodontology* 80 (2): 210–218.

Del Valle, Heather B., Ann L. Yaktine, Christine L. Taylor, and A. Catharine Ross. 2011. *Dietary reference intakes for calcium and vitamin D.* Washington, USA: National Academies Press.

Dudek, Karolina, and Tomasz Goryczka. 2016. Electrophoretic deposition and characterization of thin hydroxyapatite coatings formed on the surface of NiTi shape memory alloy. *Ceramics International* 42 (16): 19124–19132.

Eckardt, Ingo, and Hans-Joachim Hein. 2001. Quantitative measurements of the mechanical properties of human bone tissues by scanning acoustic microscopy. *Annals of Biomedical Engineering* 29 (12): 1043–1047.

Eliaz, Noam, and Noah Metoki. 2017. Calcium phosphate bioceramics: A review of their history, structure, properties, coating technologies and biomedical applications. *Materials* 10 (4): 334.

Elmay, W., F. Prima, Thierry Gloriant, et al. 2013. Effects of thermomechanical process on the microstructure and mechanical properties of a fully martensitic titanium-based biomedical alloy. *Journal of the Mechanical Behavior of Biomedical Materials* 18: 47–56.

Ford, Catherine M., and Tony M. Keaveny. 1996. The dependence of shear failure properties of trabecular bone on apparent density and trabecular orientation. *Journal of Biomechanics* 29 (10): 1309–1317.

Frost, Harold M. 1998. From Wolff's law to the mechanostat: A new "face" of physiology. *Journal of Orthopaedic Science* 3 (5): 282–286.

Fyhrie, David P., and J. H. Kimura. 1999. Cancellous bone biomechanics. *Journal of Biomechanics* 32 (11): 1139–1148.

Ge, Zigang, Lishan Wang, Boon Chin Heng, et al. 2009. Proliferation and differentiation of human osteoblasts within 3D printed poly-lactic-co-glycolic acid scaffolds. *Journal of Biomaterials Applications* 23 (6): 533–547.

Gepreel, Mohamed Abdel-Hady, and Mitsuo Niinomi. 2013. Biocompatibility of Ti-alloys for long-term implantation. *Journal of the Mechanical Behavior of Biomedical Materials* 20: 407–415.

Gilissen, R., J. P. Erauw, A. Smolders, E. Vanswijgenhoven, and J. Luyten. 2000. Gelcasting, a near net shape technique. *Materials & Design* 21 (4): 251–257.

Guo, X. Edward. 2001. Mechanical properties of cortical bone and cancellous bone tissue. *Bone Mechanics Handbook* 10: 1–23.

Han, Siyuan, Yuexian Liu, Xin Nie, et al. 2012. Efficient delivery of antitumor drug to the nuclei of tumor cells by amphiphilic biodegradable poly (L-aspartic acid-co-lactic acid)/DPPE co-polymer nanoparticles. *Small* 8 (10): 1596–1606.

Han, Wei, Xian-ming Meng, Jun-bao Zhang, and Jie Zhao. 2012. Elastic modulus of 304 stainless steel coating by cold gas dynamic spraying. *Journal of Iron and Steel Research International* 19 (3): 73–78.

Hermawan, H., H. Alamdari, D. Mantovani, and Dominique Dube. 2008. Iron–manganese: New class of metallic degradable biomaterials prepared by powder metallurgy. *Powder Metallurgy* 51 (1): 38–45.

Huang, Tao, Jian Cheng, Dong Bian, and Yufeng Zheng. 2016. Fe–Au and Fe–Ag composites as candidates for biodegradable stent materials. *Journal of Biomedical Materials Research Part B: Applied Biomaterials* 104 (2): 225–240.

Huang, Yun, Yonghao Pan, Weiwei Wang, Long Jiang, and Yi Dan. 2019. Synthesis and properties of partially biodegradable fluorinated polyacrylate: Poly (l-lactide)-co-poly (hexafluorobutyl acrylate) copolymer. *Materials & Design* 162: 285–292.

Huiskes, Rik, Harrie Weinans, and Bert Van Rietbergen. 1992. The relationship between stress shielding and bone resorption around total hip stems and the effects of flexible materials. *Clinical Orthopaedics and Related Research* 124–134.

Hussain, O. T., S. Sah, and A. J. Sidebottom. 2014. Prospective comparison study of one-year outcomes for all titanium total temporomandibular joint replacements in patients allergic to metal and cobalt–chromium replacement joints in patients not allergic to metal. *British Journal of Oral and Maxillofacial Surgery* 52 (1): 34–37.

Hutmacher, Dietmar W, Michael Sittinger, and Makarand V. Risbud. 2004. Scaffold-based tissue engineering: Rationale for computer-aided design and solid free-form fabrication systems. *TRENDS in Biotechnology* 22 (7): 354–362.

Jakubowicz, J., G. Adamek, and M. Dewidar. 2013. Titanium foam made with saccharose as a space holder. *Journal of Porous Materials* 20 (5): 1137–1141.

Jenei, Péter, Hyelim Choi, Adrián Tóth, Heeman Choe, and Jenő Gubicza. 2016. Mechanical behavior and microstructure of compressed Ti foams synthesized via freeze casting. *Journal of the Mechanical Behavior of Biomedical Materials* 63: 407–416.

Jha, Nidhi, D. P. Mondal, J. Dutta Majumdar, Anshul Badkul, A. K. Jha, and A. K. Khare. 2013. Highly porous open cell Ti-foam using NaCl as temporary space holder through powder metallurgy route. *Materials & Design* 47: 810–819.

Jung, KiMin, SooHoon Ahn, YoungJun Kim, SeKwon Oh, Won-Hee Ryu, and HyukSang Kwon. 2018. Alloy design employing high Cr concentrations for Mo-free stainless steels with enhanced corrosion resistance. *Corrosion Science* 140: 61–72.

Kadkhodapour, J., H. Montazerian, A. Ch Darabi, et al. 2015. Failure mechanisms of additively manufactured porous biomaterials: Effects of porosity and type of unit cell. *Journal of the Mechanical Behavior of Biomedical Materials* 50: 180–191.

Kang, Min-Ho, Hyun Lee, Tae-Sik Jang, et al. 2019. Biomimetic porous Mg with tunable mechanical properties and biodegradation rates for bone regeneration. *Acta Biomaterialia* 84: 453–467.

Kao, Chia-Tze, Chi-Chang Lin, Yi-Wen Chen, Chia-Hung Yeh, Hsin-Yuan Fang, and Ming-You Shie. 2015. Poly (dopamine) coating of 3D printed poly (lactic acid) scaffolds for bone tissue engineering. *Materials Science and Engineering: C* 56: 165–173.

Kim, Hwan D., Sivashanmugam Amirthalingam, Seunghyun L. Kim, Seunghun S. Lee, Jayakumar Rangasamy, and Nathaniel S. Hwang. 2017. Biomimetic materials and fabrication approaches for bone tissue engineering. *Advanced Healthcare Materials* 6 (23): 1700612.

Kirkland, Nicholas Travis, and Nick Birbilis. 2014. *Magnesium biomaterials: design, testing, and best practice.* Cham, Switzerland: Springer.

Koike, M., and H. Fujii. 2001. In vitro assessment of corrosive properties of titanium as a biomaterial. *Journal of Oral Rehabilitation* 28 (6): 540–548.

Kubásek, J., D. Vojtěch, E. Jablonská, I. Pospíšilová, J. Lipov, and T. Ruml. 2016. Structure, mechanical characteristics and in vitro degradation, cytotoxicity, genotoxicity and mutagenicity of novel biodegradable Zn–Mg alloys. *Materials Science and Engineering: C* 58: 24–35.

Kuroda, Daisuke, Sachiko Hiromoto, Takao Hanawa, and Yasuyuki Katada. 2002. Corrosion behavior of nickel-free high nitrogen austenitic stainless steel in simulated biological environments. *Materials Transactions* 43 (12): 3100–3104.

Lasprilla, Astrid J. R., Guillermo A. R. Martinez, Betânia H. Lunelli, André L. Jardini, and Rubens Maciel Filho. 2012. Poly-lactic acid synthesis for application in biomedical devices—A review. *Biotechnology Advances* 30 (1): 321–328.

Lee, Hyub, Chin Huat Joel Lim, Mun Ji Low, Nicholas Tham, Vadakke Matham Murukeshan, and Young-Jin Kim. 2017. Lasers in additive manufacturing: A review. *International Journal of Precision Engineering and Manufacturing-Green Technology* 4 (3): 307–322.

Leukers, B., H. Gülkan, S. H. Irsen, et al. 2005. Biocompatibility of ceramic scaffolds for bone replacement made by 3D printing. *Materialwissenschaft und Werkstofftechnik:*

*Entwicklung, Fertigung, Prüfung, Eigenschaften und Anwendungen technischer Werkstoffe* 36 (12): 781–787.

Li, Huafang, Yufeng Zheng, and Ling Qin. 2014. Progress of biodegradable metals. *Progress in Natural Science: Materials International* 24 (5): 414–422.

Li, Jinyang, Si Wu, Eunkyoung Kim, et al. 2019. Electrobiofabrication: Electrically-based fabrication with biologically-derived materials. *Biofabrication* 11 (3): 032002.

Li, Y., H. Jahr, P. Pavanram, et al. 2019. Additively manufactured functionally graded biodegradable porous iron. *Acta Biomaterialia* 96: 646–661.

Liu, I-Hsin, Shih-Hsin Chang, and Hsin-Yi Lin. 2015. Chitosan-based hydrogel tissue scaffolds made by 3D plotting promotes osteoblast proliferation and mineralization. *Biomedical Materials* 10 (3): 035004.

Long, Marc, and H. J. Rack. 1998. Titanium alloys in total joint replacement—a materials science perspective. *Biomaterials* 19 (18): 1621–1639.

Michalska, Joanna, Maciej Sowa, Robert P Socha, Wojciech Simka, and Beata Cwalina. 2017. The influence of Desulfovibrio desulfuricans bacteria on a Ni-Ti alloy: Electrochemical behavior and surface analysis. *Electrochimica Acta* 249: 135–144.

Mondrinos, Mark J., Robert Dembzynski, Lin Lu, et al. 2006. Porogen-based solid freeform fabrication of polycaprolactone–calcium phosphate scaffolds for tissue engineering. *Biomaterials* 27 (25): 4399–4408.

Mozdzen, Laura C., Ryan Rodgers, Jessica M. Banks, Ryan C. Bailey, and Brendan A. C. Harley. 2016. Increasing the strength and bioactivity of collagen scaffolds using customizable arrays of 3D-printed polymer fibers. *Acta Biomaterialia* 33: 25–33.

Nandakumar, Anandkumar, Hugo Fernandes, Jan de Boer, Lorenzo Moroni, Pamela Habibovic, and Clemens A. van Blitterswijk. 2010. Fabrication of bioactive composite scaffolds by electrospinning for bone regeneration. *Macromolecular Bioscience* 10 (11): 1365–1373.

Nocera, Aden Díaz, Romina Comín, Nancy Alicia Salvatierra, and Mariana Paula Cid. 2018. Development of 3D printed fibrillar collagen scaffold for tissue engineering. *Biomedical Microdevices* 20 (2): 26.

Nouri, Alireza, Peter D. Hodgson, and Cui'e Wen. 2010. *Biomimetic porous titanium scaffolds for orthopaedic and dental applications.* London, UK: Intech.

Öhman, Caroline, Massimiliano Baleani, and Carla Pani, et al. 2011. Compressive behaviour of child and adult cortical bone. *Bone* 49 (4):769–776.

Oriňáková, Renáta, Andrej Oriňák, and Lucia M Bučková, et al. 2013. Iron based degradable foam structures for potential orthopedic applications. *International Journal of Electrochemical Science* 8: 12451–12465.

Parthasarathy, Jayanthi, Binil Starly, and Shivakumar Raman. 2011. A design for the additive manufacture of functionally graded porous structures with tailored mechanical properties for biomedical applications. *Journal of Manufacturing Processes* 13 (2): 160–170.

Quelch, Kaylene J., R. A. Melick, Patricia J. Bingham, and Susan M. Mercuri. 1983. Chemical composition of human bone. *Archives of Oral Biology* 28 (8): 665–674.

Rae, T. 1975. A study on the effects of particulate metals of orthopaedic interest on murine macrophages in vitro. *The Journal of Bone and Joint Surgery. British Volume* 57 (4): 444–450.

Rae, T. 1981. The toxicity of metals used in orthopaedic prostheses. An experimental study using cultured human synovial fibroblasts. *The Journal of Bone and Joint Surgery. British Volume* 63 (3): 435–440.

Rho, Jae Young, Richard B. Ashman, and Charles H. Turner. 1993. Young's modulus of trabecular and cortical bone material: Ultrasonic and microtensile measurements. *Journal of Biomechanics* 26 (2): 111–119.

Rho, Jae-Young, Liisa Kuhn-Spearing, and Peter Zioupos. 1998. Mechanical properties and the hierarchical structure of bone. *Medical Engineering & Ohysics* 20 (2): 92–102.

Sadat-Shojai, Mehdi, Mohammad-Taghi Khorasani, Ehsan Dinpanah-Khoshdargi, and Ahmad Jamshidi. 2013. Synthesis methods for nanosized hydroxyapatite with diverse structures. *Acta Biomaterialia* 9 (8): 7591–7621.

Saranya, N., S. Saravanan, A. Moorthi, B. Ramyakrishna, and N. Selvamurugan. 2011. Enhanced osteoblast adhesion on polymeric nano-scaffolds for bone tissue engineering. *Journal of Biomedical Nanotechnology* 7 (2): 238–244.

Sepulveda, P., J. G. P. Binner, S. O. Rogero, O. Z. Higa, and J. C. Bressiani. 2000. Production of porous hydroxyapatite by the gel-casting of foams and cytotoxic evaluation. *Journal of Biomedical Materials Research: An Official Journal of The Society for Biomaterials and The Japanese Society for Biomaterials* 50 (1): 27–34.

Sharma, M., G. K. Gupta, O. P. Modi, B. K. Prasad, and Anil K. Gupta. 2011. Titanium foam through powder metallurgy route using acicular urea particles as space holder. *Materials Letters* 65 (21–22): 3199–3201.

Shim, V. P. W., L. M. Yang, J. F. Liu, and V. S. Lee. 2005. Characterisation of the dynamic compressive mechanical properties of cancellous bone from the human cervical spine. *International Journal of Impact Engineering* 32 (1–4): 525–540.

Shuai, Cijun, Sheng Li, Shuping Peng, Pei Feng, Yuxiao Lai, and Chengde Gao. 2019. Biodegradable metallic bone implants. *Materials Chemistry Frontiers* 3: 544–562.

Sims, Natalie A., and Jonathan H. Gooi. 2008. Bone remodeling: Multiple cellular interactions required for coupling of bone formation and resorption. *Seminars in Cell & Developmental Biology* 19 (5): 444–451.

Singh, Pradeep, Amit Abhash, Prashant Nair, Anup Khare, I. B. Singh, and D. P. Mondal. 2019. Effect of space holder size on microstructure, deformation and corrosion response of $Ti_4Al_4Co$ (wt%) alloy foam. *Applied Innovative Research (AIR)* 1 (1): 41–47.

Singh, Pradeep, Amit Abhash, B. N. Yadav, M. Shafeeq, I. B. Singh, and D. P. Mondal. 2019. Effect of milling time on powder characteristics and mechanical performance of Ti4wt% Al alloy. *Powder Technology* 342: 275–287.

Singh, Pradeep, I. B. Singh, and D. P. Mondal. 2019. A comparative study on compressive deformation and corrosion behaviour of heat treated Ti4wt% Al foam of different porosity made of milled and unmilled powders. *Materials Science and Engineering: C* 98: 918–929.

Snyder, Susan M., and Erich Schneider. 1991. Estimation of mechanical properties of cortical bone by computed tomography. *Journal of Orthopaedic Research* 9 (3): 422–431.

Speck, Karen M., and Anna C. Fraker. 1980. Anodic polarization behavior of Ti-Ni and Ti-6A 1-4 V in simulated physiological solutions. *Journal of Dental Research* 59 (10): 1590–1595.

Staiger, Mark P., Alexis M. Pietak, Jerawala Huadmai, and George Dias. 2006. Magnesium and its alloys as orthopedic biomaterials: A review. *Biomaterials* 27 (9): 1728–1734.

Sun, Xiaoyu, Chun Xu, Gang Wu, Qingsong Ye, and Changning Wang. 2017. Poly (lactic-co-glycolic acid): Applications and future prospects for periodontal tissue regeneration. *Polymers* 9 (6): 189.

Tang, Zhaohui, Chaoliang He, Huayu Tian, et al. 2016. Polymeric nanostructured materials for biomedical applications. *Progress in Polymer Science* 60: 86–128.

Torres, Yadir, Sheila Lascano, Jorge Bris, Juan Pavón, and José A. Rodriguez. 2014. Development of porous titanium for biomedical applications: A comparison between loose sintering and space-holder techniques. *Materials Science and Engineering: C* 37: 148–155.

Vella, Joseph B., Ryan P. Trombetta, Michael D. Hoffman, Jason Inzana, Hani Awad, and Danielle S. W. Benoit. 2018. Three dimensional printed calcium phosphate and poly (caprolactone) composites with improved mechanical properties and preserved microstructure. *Journal of Biomedical Materials Research Part A* 106 (3): 663–672.

Wang, Hao, Yuri Estrin, Huameng Fu, Guangling Song, and Zuzana Zúberová. 2007. The effect of pre-processing and grain structure on the bio-corrosion and fatigue resistance of magnesium alloy AZ31. *Advanced Engineering Materials* 9 (11): 967–972.

Wang, Huifang, Yamei Leng, and Yuping Gong. 2018. Bone marrow fat and hematopoiesis. *Frontiers in Endocrinology* 9: 694.

Wang, Mayao, Simin Li, Annika vom Scheidt, Mahan Qwamizadeh, Björn Busse, and Vadim V. Silberschmidt. 2020. Numerical study of crack initiation and growth in human cortical bone: Effect of micro-morphology. *Engineering Fracture Mechanics* 107051.

Wegener, Bernd, Birte Sievers, Sandra Utzschneider, et al. 2011. Microstructure, cytotoxicity and corrosion of powder-metallurgical iron alloys for biodegradable bone replacement materials. *Materials Science and Engineering: B* 176 (20): 1789–1796.

White, Tim D., Michael T. Black, and Pieter A. Folkens. 2011. *Human osteology.* Burlington, MA, USA: Academic Press.

Williams, Jessica M., Adebisi Adewunmi, Rachel M. Schek, et al. 2005. Bone tissue engineering using polycaprolactone scaffolds fabricated via selective laser sintering. *Biomaterials* 26 (23): 4817–4827.

Witte, Frank, V. Kaese, H. Haferkamp, et al. 2005. In vivo corrosion of four magnesium alloys and the associated bone response. *Biomaterials* 26 (17): 3557–3563.

Xin, Yunchang, Chenglong Liu, Xinmeng Zhang, Guoyi Tang, Xiubo Tian, and Paul K. Chu. 2007. Corrosion behavior of biomedical AZ91 magnesium alloy in simulated body fluids. *Journal of Materials Research* 22 (7): 2004–2011.

Yan, Chunze, Liang Hao, Ahmed Hussein, and Philippe Young. 2015. Ti–6Al–4V triply periodic minimal surface structures for bone implants fabricated via selective laser melting. *Journal of the Mechanical Behavior of Biomedical Materials* 51: 61–73.

Yang, Shoufeng, Kah-Fai Leong, Zhaohui Du, and Chee-Kai Chua. 2001. The design of scaffolds for use in tissue engineering. Part I. Traditional factors. *Tissue Engineering* 7 (6): 679–689.

Yang, Kun, Jian Wang, Liang Jia, Guangyu Yang, Huiping Tang, and Yuanyuan Li. 2019. Additive manufacturing of Ti-6Al-4V lattice structures with high structural integrity under large compressive deformation. *Journal of Materials Science & Technology* 35 (2): 303–308.

Yang, Shoufeng, Kah-Fai Leong, Zhaohui Du, and Chee-Kai Chua. 2002. The design of scaffolds for use in tissue engineering. Part II. Rapid prototyping techniques. *Tissue Engineering* 8 (1): 1–11.

Zhang, Shucai, Huabing Li, Zhouhua Jiang, et al. 2019. Effects of Cr and Mo on precipitation behavior and associated intergranular corrosion susceptibility of superaustenitic stainless steel S32654. *Materials Characterization* 152: 141–150.

Zubair, M., S. Sandlöbes, M. A. Wollenweber, et al. 2019. On the role of Laves phases on the mechanical properties of Mg-Al-Ca alloys. *Materials Science and Engineering: A* 756: 272–283.

# 10 Hybrid Optimisation for Supply Chain Management: A Case of Supplier Selection by CRITIC, ARAS and TOPSIS Techniques

*Josy George, Pushkal Badoniya, and Dr. J. Francis Xavier*

## CONTENTS

## 10.1  INTRODUCTION

Supplier selection and evaluation are painstaking elements for SCM, which grant a supplier-manufacturing firm relationship to establish a long-term commitment to each other. Designing novel goods and services, gaining possession of raw material, processing the same components to produce valuable products in the final stage and

**161**

finally making deliveries to the demand creators are the components that a supply chain is comprised of, and that proves that supply chains are tangible in both service and manufacturing organisations (Chen 2011). The issue transforms to a pivotal phase in manufacturing sectors where a large amount of time and funds are disbursed for acquiring purposes. Top-notch suppliers make organisations capable of achieving praiseworthy production performance which, in turn, keeps the organisation solvent by providing it with optimum profits. Successfully accomplished supply chain management provides perks such as novel efficiency, increased profits, decrement in price and increment in consonance, thus allowing the organisation to deal with demand in an enhanced way, hold sufficient inventory in storage, manage interruptions, restrain costs and optimally reciprocate for the demand (Abo-Sinna and Amer 2005).

Multi-criteria decision making (MCDM) alludes to choosing the foremost option from a limited number of decisive options in terms of various and mostly dissonant standards. The MCDM methodology is quite popular among businesses, industries and manufacturing sectors, etc. Multiple criteria decision making (MCDM) techniques mostly require the administrator to evaluate alternatives while being mindful of the standards for the decision plus implementing it with significant values (Hwang and Yoon, 1981). A prime alternative is then picked and hinged on the assigned weights to the criteria and then ranked to the alternatives. Technical developments such as computers have alleviated and made the recondite methods easier to comprehend for users, thus enabling them to be applied in a myriad of decision-making fields such as economy and management, among others (Jee and Kang 2000). According to research, a number of methods is derived for weightage and rank calculation. From various multi-criteria methodologies, MAXMIN, MAXMAX, SAW, AHP, TOPSIS, SMART, COPRAS, MOORA and ELECTRE are the most popularly used. The three salient procedures in these methodologies are data gathering, processing and construal (Jayant and Singh 2015). The first step is data acquisition based on the type of problem, for example, material selection, supplier selection, software selection, etc. (Weber, Current, and Benton 1991). The data is different, because material selection parameters of the material are required. When choosing the supplier, the relative data of supplier determination criteria are considered.

A wide range of decision support techniques has been applied to solve the multi-criteria decision making (MCDM) problem pertaining to supplier selection. The major complexity is the selection of a sustainable supplier (Ho, Xu, and Dey 2010). MCDM techniques are mainly used to select the best alternative considering decision criteria (Dickson 1966). The major MCDM techniques that have been used in various past work include analytical hierarchy process (AHP), data envelopment analysis (DEA), analytic network process (ANP), ELECTRE (Elimination Et Choix Traduisant la REalite – ELimination and Choice Expressing the Reality), fuzzy approaches, PROMETHEE (Preference Ranking Organisation Method for Enrichment Evaluation), artificial neural network (ANN)-based approaches, simple multi-attribute rating technique (SMART), and others. Based on the networking concept a hybrid model of Bayesian Networks (BN) and Total Cost of Ownership

(TCO) methods explained for supplier selection with the condition of limited and uncertain information regarding the supplier (Dogan and Aydin 2011).

## 10.2 OPTIMISATION

### 10.2.1 CRITIC Method

The criteria of importance through intercriteria correlation (CRITIC) method refers to the standard deviation formulated by Diakoulaki et al., which adopts correlation analysis to determine values of a particular standard. It is based on the systematic scrutiny of the evaluation matrix for gathering information comprised in the determination method (Ighravwe and Babatunde 2018). It can be surmised that objective weights are obtained by measuring innate data associated with specific determination criteria. In this method, the evaluation procedure for standard weights consists of the standard deviation of the rule and its interaction with different correlations.

For instance, the initial decision matrix, $X = [x_{ij}]_{m*n}$ comprising of m options and n criteria, in which xij symbolises the performance measurement of i[th] alternative or option correlated to j[th] criterion. The following acronyms are used to determine the weight of j[th] criterion: $c_j$ is the quantity of information contained in the j[th] criterion, $\sigma_j$ is the standard deviation of the j[th] criterion and $\rho_{ij}$ is the correlation coefficient between the j[th] and k[th] criteria. Furthermore, the step-wise method for solving the CRITIC method is as follows:

Step 1. Normalising the initial decision matrix using the succeeding equations for beneficial criteria and non-beneficial (cost) criteria:

$$r_{ij} = \frac{x_{ij} - x_j^{min}}{x_j^{max} - x_j^{min}} \quad i = 1, \ldots, m; j = 1, \ldots, n$$

$$r_{ij} = \frac{x_j^{max} - x_{ij}}{x_j^{max} - x_j^{min}} \quad i = 1, \ldots, m; j = 1, \ldots,$$

Every vector has a standard deviation which represents the level of deviation of the variant values for given criteria of a mean.

Step 2. Criteria pairs correlation calculation

$$\rho_{jk} = \frac{\sum_{i=1}^{m} (r_{ij} - \overline{r_j})(r_{ik} - \overline{r_j}_k)}{\sqrt{\sum_{i=1}^{m} (r_{ij} - \overline{r_j})^2 \sum_{i=1}^{m} (r_{ik} - \overline{r_j}_k)^2}}$$

Step 3. Criteria weights calculation

$$c_j = \sigma_j \sum_{k=1}^{n} (1 - \rho_{jk})$$

$$w_j = \frac{c_j}{\sum_{k=1}^{n} c_k}$$

where: i = 1, 2, ..., m; j, k = 1, 2, ..., n.

## 10.2.2  ARAS Method

A classic MCDM issue deals with the job of ordering limited instances of decision substitutes, each of which is clearly delineated in terms of distinct decision standards which are dealt with concurrently. In the study, this methodology is applied for the performance evaluation of vendors for any manufacturing firm (Turskis and Zavadskas 2010). According to the ARAS method, its primary function is in discovering the sophisticated corresponding efficiency of a possible substitute that is directly proportional to the comparative consequence of values and weights of the front-running criteria that are taken into account in an issue (Dadelo et al. 2012).

The process involved in determining the solutions of such issues with the aid of this method, in cases when the MCDM problem includes beneficial criterion and non-beneficial criterion, can be accurately delineated with the succeeding procedure (Chatterjee and Chakraborty 2013).

Step 1: First, the related decision/evaluation matrix is formulated. In any MCDM problem (discrete optimisation problem), the relevant data is illustrated with the aid of a decision matrix showing preferences from pragmatic alternatives ordered by n criteria (attributes).

$$X = \begin{bmatrix} x_{12} & x_{12} & \cdots & x_{1n} \\ x_{21} & x_{22} & \cdots & x_{2n} \\ \cdot & \cdot & \cdots & \cdot \\ \cdot & \cdot & \cdots & \cdot \\ \cdot & \cdot & \cdots & \cdot \\ x_{m1} & x_{m2} & \cdots & x_{mn} \end{bmatrix}$$

where m is the number of alternatives, n is the number of criteria describing each alternative and $x_{ij}$ is the value representing the performance of $i^{th}$ alternative w.r.t $j^{th}$ criterion.

Step 2: Next, the optimal value of each criterion must be determined. Let $x_{0j}$ be the optimal value of $j^{th}$ criterion. If the optimal value of $j^{th}$ criterion is known, then

$$x_{0j} = \max x_{ij} \quad \text{for beneficial criterion}$$

$$x_{0j} = \min x_{ij} \quad \text{for non} - \text{beneficial criterion}$$

At this point, considering the optimal values of all the included criteria, the original decision matrix is reformulated as below:

$$X = \begin{bmatrix} x_{01} & x_{0j} & \cdots & x_{0n} \\ \cdot & \cdot & \cdots & \cdot \\ x_{i1} & x_{ij} & \cdots & x_{in} \\ \cdot & \cdot & \cdots & \cdot \\ \cdot & \cdot & \cdots & \cdot \\ x_{m1} & x_{mj} & \cdots & x_{mn} \end{bmatrix}$$

Step 3: In this step, all of the initial criteria values are normalised while employing the following equations.

$$\text{For beneficial criteria,} \quad r_{ij} = \frac{x_{ij}}{\sum_{i=0}^{m} x_{ij}}$$

$$\text{For non} - \text{beneficial criteria,} \quad r_{ij}^* = \frac{1}{x_{ij}}, \; r_{ij} = \frac{r_{ij}^*}{\sum_{i=0}^{m} r_{ij}^*}$$

Step 4: From the normalised decision matrix, the corresponding weighted normalised decision matrix is developed using the following equation:

$$v_{ij} = w_j * r_{ij}, \quad i = 1, 2, 3 \ldots, m$$

Where, $w_j$ is the weight of $j^{th}$ criterion and $r_{ij}$ is the normalised performance of $i^{th}$ alternative with respect to $j^{th}$ criterion.

Step 5: At this stage, the optimality function value is determined.

$$S_i = \sum_{j=1}^{n} v_{ij}, \quad i = 0, 1, 2 \; 3 \ldots, m$$

where $S_i$ is the value of optimality function for $i^{th}$ alternative. The greatest magnitude of Si always signifies the best alternative, whereas, the lowest Si value identifies the least preferred alternative. Taking into account the computational process of the ARAS method (Kutut, Zavadskas, and Lazauskas 2014), it can be revealed that the optimality function $S_i$ is commensurately related to $x_{ij}$ values and weights $w_j$ of the considered criteria and comparative effect of the same on the absolute output. The priorities of the substitutes can thus be evaluated on the basis of $S_i$ values. As a result, the ranking of the decision alternatives using $S_i$ values becomes much simpler and hassle-free.

Step 6: The degree of the substitute's usefulness is evaluated by comparing it with an alteration, which is often considered as the hypothetically most optimal value ($S_0$). The utility degree $U_i$ of $i^{th}$ alternative can be calculated employing the following equation:

$$U_i = \frac{S_i}{S_0}, \quad i = 1, 2, \ldots, m$$

It is apparent that the evaluated value of $U_i$ lies in the interval of [0, 1] and can be marshalled in an increasing order to be able to provide a categorical placing list of the alternatives that are taken into account (Esbouei and Ghadikolaei 2013). The relative productivity of the achievable alternatives can likewise be resolved by the utility function values.

## 10.2.3 TOPSIS METHOD

A multi-criteria decision making (MCDM) methodology aids the decision-makers in determining top-notch substitutes. TOPSIS is a frequently used methodology of MADM models or the multi-attribute decision-making model (Mousavi-Nasab and Sotoudeh-Anvari 2017). The Technique for Order Preference by Similarity to Ideal Solution (TOPSIS) was developed by Hwang and Yoon 1981, which is a procedure of multi-criteria decision analysis. TOPSIS dialectic is comprehensible and sensible, which picks a substitute, bears the smallest geometric distance from the positive best substitute and juxtaposes a lot of options or alternatives (Boran et al. 2009) for recognising weights for every single criterion and normalises the scores for each standard. Next, it assesses the geometric path between evaluated and ideal substitutes for providing the optimal score for every single standard. TOPSIS methodology aids in determining the perfect supplier along with a distinct yet limited number of criteria (Iç 2012).

Step 1: Identify the structure of matrix

$$X = \begin{bmatrix} x_{12} & x_{12} & \cdots & x_{1n} \\ x_{21} & x_{22} & \cdots & x_{2n} \\ \cdot & \cdot & \cdot & \cdot \\ \cdot & \cdot & \cdot & \cdot \\ x_{m1} & x_{m2} & \cdots & x_{mn} \end{bmatrix}$$

Step 2: Calculate the normalised matrix X by using the following formula:

$$r_{ij} = \frac{x_{ij}}{\sqrt{\sum_{j=1}^{J} x_{ij}^2}}$$

Step 3: Develop the weighted normalised decision matrix by multiplying:

$$V_{ij} = w_{ij} * r_{ij}$$

Step 4: Determine the positive ideal solution and negative ideal solution:

$$A^* = \{ (max\ v_{ij} | j \in J), (min\ v_{ij} | j \in J') \}$$

$$A^- = \{ (min\ v_{ij} | j \in J), (max\ v_{ij} | j \in J') \}$$

Step 5: Find the separation measure value with the equations below:

$$S_i^* = \sqrt{\sum_{j=1}^{n} (v_{ij} - v_j^*)^2}$$

$$S_i^- = \sqrt{\sum_{j=1}^{n} (v_{ij} - v_j^-)^2}$$

Step 6: Determine the relative closeness to the ideal solution by:

$$C_i^* = S_i^- / S_i^* + S_i^- \quad 0 \le C_i^* \le 1$$

Step 7: Identify the total score and select the closest value to 1, which is the most suitable alternative.

## 10.3 CASE STUDY

A manufacturing organisation intends on finding the most ideal of their suppliers for a supply among ten suppliers based on four important criteria. The application of the MCDM approach is considered in solving this type of problem, it determines the most suitable supplier who will meet the requirements at optimum level and according to the criteria values set by management. Distinctive standards included in the decision making procedure of which criterion to include are: Quality of product, price/cost of the product, timely delivery and service. In this case, the criteria, price/cost is not profitable but features concerned with other criteria are profitable. Table 10.1 lists down suppliers and their corresponding ratings according to different criteria.

**TABLE 10.1**
**Decision Matrix**

|  | Quality | Price | Delivery | Service |
|---|---|---|---|---|
| Supplier 1 | 7.6 | 390 | 11 | 46 |
| Supplier 2 | 5.5 | 360 | 11 | 32 |
| Supplier 3 | 5.3 | 290 | 11 | 32 |
| Supplier 4 | 5.7 | 270 | 9 | 37 |
| Supplier 5 | 4.2 | 240 | 8 | 38 |
| Supplier 6 | 4.4 | 260 | 8 | 38 |
| Supplier 7 | 3.9 | 270 | 5 | 42 |
| Supplier 8 | 7.9 | 400 | 6 | 44 |
| Supplier 9 | 8.1 | 380 | 6 | 44 |
| Supplier 10 | 4.5 | 320 | 7 | 46 |
| Supplier 11 | 5.7 | 320 | 11 | 48 |
| Supplier 12 | 5.2 | 310 | 11 | 48 |
| Supplier 13 | 7.1 | 280 | 12 | 49 |
| Supplier 14 | 6.9 | 250 | 10 | 50 |
| Criteria Type | Max | Min | Max | Max |

**TABLE 10.2**

**Normalised Matrix for the CRITIC Method**

|              | Quality | Price  | Delivery | Service |
|--------------|---------|--------|----------|---------|
| Supplier 1   | 0.8810  | 0.0625 | 0.8571   | 0.7778  |
| Supplier 2   | 0.3810  | 0.2500 | 0.8571   | 0.0000  |
| Supplier 3   | 0.3333  | 0.6875 | 0.8571   | 0.0000  |
| Supplier 4   | 0.4286  | 0.8125 | 0.5714   | 0.2778  |
| Supplier 5   | 0.0714  | 1.0000 | 0.4286   | 0.3333  |
| Supplier 6   | 0.1190  | 0.8750 | 0.4286   | 0.3333  |
| Supplier 7   | 0.0000  | 0.8125 | 0.0000   | 0.5556  |
| Supplier 8   | 0.9524  | 0.0000 | 0.1429   | 0.6667  |
| Supplier 9   | 1.0000  | 0.1250 | 0.1429   | 0.6667  |
| Supplier 10  | 0.1429  | 0.5000 | 0.2857   | 0.7778  |
| Supplier 11  | 0.4286  | 0.5000 | 0.8571   | 0.8889  |
| Supplier 12  | 0.3095  | 0.5625 | 0.8571   | 0.8889  |
| Supplier 13  | 0.7619  | 0.7500 | 1.0000   | 0.9444  |
| Supplier 14  | 0.7143  | 0.9375 | 0.7143   | 1.0000  |
| Sigma ($\sigma_j$) | 0.3389 | 0.3361 | 0.3315 | 0.3374 |

### 10.3.1 IMPLEMENTATION OF THE CRITIC METHOD FOR WEIGHT CALCULATION

The CRITIC method or Criteria Importance Through Intercriteria Correlation lies in the category of interdependent methodologies. It stems from logical testing for the decision matrix which, in turn, aids in determining the data included in the criteria responsible for the calculation and identification of variants.

Step 1: Normalise the initial decision matrix and standard deviation for each criterion (Table 10.2).

Step 2: Construct a symmetric matrix for the given criteria as shown in Table 10.3, with dimensions mxm and a generic element $r_{jk}$, which is the linear correlation coefficient between the vectors $x_j$ and $x_k$.

**TABLE 10.3**

**Linear Correlation Coefficient between the Criteria**

|          | Quality | Price   | Delivery | Service |
|----------|---------|---------|----------|---------|
| Quality  | 1.0000  | −0.6351 | 0.1304   | 0.3974  |
| Price    | −0.6351 | 1.0000  | 0.0801   | −0.1013 |
| Delivery | 0.1304  | 0.0801  | 1.0000   | 0.0491  |
| Service  | 0.3974  | −0.1013 | 0.0491   | 1.0000  |

**TABLE 10.4**

**The Amount of Information and Criteria Weights**

| Measure of Conflict | Amount of Information (Cj) | Objective Weights (Wi) |
|---|---|---|
| 3.1073 | 1.0530 | 0.2577 |
| 3.6563 | 1.2290 | 0.3008 |
| 2.7403 | 0.9084 | 0.2223 |
| 2.6547 | 0.8956 | 0.2192 |
| | 4.0860 | 1.0000 |

Step 3: Determine the weightage of the criteria (Table 10.4)

Figure 10.1 represents a 2D bar chart diagram. There are four criteria, i.e. quality, price, delivery and service as represented in the x-axis. For each criterion, there are three columns that represent the values of measure of conflict, amount of information and objective weight. The comparison of different criteria values becomes easy to understand with the succeeding bar chart.

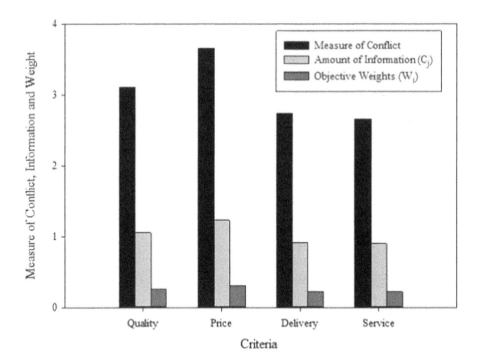

**FIGURE 10.1** Measure of Conflict, Amount of Information, Objective Weights and Criteria.

## 10.3.2  IMPLEMENTATION OF THE ARAS METHOD FOR THE RANKING OF SUPPLIERS

An exact normalisation procedure has been explained in the theory section, where the profitable features and reciprocals of a criterion are taken into account for non-profitable characteristics. Extrapolated from the given relationship in the section, evaluation of the degree of utility for the entire number of suppliers is done (Table 10.5).

The degree of utility values of the alternatives varies between 0 and 100% and the highest utility value is considered as the best preference for supply.

## 10.3.3  IMPLEMENTATION OF THE TOPSIS METHOD FOR THE RANKING OF SUPPLIERS

In the TOPSIS method, the calculation standard can be divided into two classes, benefit and cost attributes. The concept is that the selected substitute should have the shortest path from the PIS and the farthest from the NIS in dealing with a multiple criteria decision-making issue. The evaluation procedures of this methodology are followed by normalising the decision matrix and then determining the relative proximity coefficient for each substitute (Table 10.6).

## 10.3.4  COMPARISON BETWEEN THE ARAS AND TOPSIS METHODS

For the given problem, the weights obtained by the CRITIC method are listed with the ARAS and TOPSIS methods. In short, the hybrid models of CRITIC-ARAS and

## TABLE 10.5
## Si, Ui Values and Ranking

|            | Si     | Ui     | Rank |
|------------|--------|--------|------|
| **Supplier 0** | 0.0843 | 1.0000 |      |
| Supplier 1 | 0.0704 | 0.8347 | 3    |
| Supplier 2 | 0.0609 | 0.7220 | 12   |
| Supplier 3 | 0.0643 | 0.7625 | 7    |
| Supplier 4 | 0.0654 | 0.7761 | 6    |
| Supplier 5 | 0.0626 | 0.7428 | 9    |
| Supplier 6 | 0.0613 | 0.7270 | 11   |
| Supplier 7 | 0.0555 | 0.6588 | 14   |
| Supplier 8 | 0.0621 | 0.7367 | 10   |
| Supplier 9 | 0.0635 | 0.7528 | 8    |
| Supplier 10 | 0.0584 | 0.6927 | 13   |
| Supplier 11 | 0.0690 | 0.8179 | 4    |
| Supplier 12 | 0.0681 | 0.8080 | 5    |
| Supplier 13 | 0.0776 | 0.9200 | 1    |
| Supplier 14 | 0.0767 | 0.9093 | 2    |

**TABLE 10.6**

**Separation Measures, Relative Closeness Coefficient and Ranking Order**

|             | S⁺    | S⁻    | C*    | Rank |
|-------------|-------|-------|-------|------|
| Supplier 1  | .0445 | .0688 | .6069 | 4    |
| Supplier 2  | .0557 | .0508 | .4772 | 9    |
| Supplier 3  | .0479 | .0579 | .5469 | 7    |
| Supplier 4  | .0438 | .0537 | .5507 | 6    |
| Supplier 5  | .0609 | .0521 | .4608 | 10   |
| Supplier 6  | .0592 | .0474 | .4447 | 12   |
| Supplier 7  | .0765 | .0403 | .3449 | 14   |
| Supplier 8  | .0650 | .0547 | .4569 | 11   |
| Supplier 9  | .0611 | .0574 | .4844 | 8    |
| Supplier 10 | .0639 | .0357 | .3583 | 13   |
| Supplier 11 | .0391 | .0608 | .6086 | 3    |
| Supplier 12 | .0428 | .0598 | .5828 | 5    |
| Supplier 13 | .0172 | .0794 | .8218 | 1    |
| Supplier 14 | .0216 | .0741 | .7739 | 2    |

CRITIC-TOPSIS models are prepared and demonstrated. The comparison of both the methods CRITIC-ARAS and CRITIC-TOPSIS is illustrated in Table 10.7 and Figure 10.2.

## 10.4 RESULT AND CONCLUSION

Multi-criteria decision making is popularly implemented for tasks like the determination of solutions to decision-making problems, in which several factors are present in the process of extracting the optimal answer, and distinct techniques are put in place to solve recondite issues. The objective is to determine the perfect supplier capable of fulfilling each and every standard of the manufacturer in the most efficient situation. Many methodologies are present in the MCDM approach, in which AHP, FAHP, VIKOR, COPRAS, ARAS and TOPSIS are the frequently used methodologies for these problems and in comparison with other techniques and available methods. The proposed research work has concentrated on issues and complexities in applying the ARAS and TOPSIS methods to real problems, such as supplier selection problems in supply chain management. The general CRITIC method is hybridised with the ARAS and TOPSIS methods, A numerical illustration is presented for utilisation and comparison studies with the ARAS and TOPSIS methods for supplier selection problems together with the entropy method.

According to the results, the ARAS and TOPSIS methods provide approximately same solutions at the top of the ranking list. Table 10.7 and Figure 10.2 show the respective results of ARAS and TOPSIS combined with CRITIC. Both methods identified that supplier 13 is the most suitable supplier for the firm, but when the

**TABLE 10.7**

**Ranking Results Obtained Using ARAS and TOPSIS Method**

| Alternatives | ARAS Method | TOPSIS Method |
|---|---|---|
| Supplier 1 | 3 | 4 |
| Supplier 2 | 12 | 9 |
| Supplier 3 | 7 | 7 |
| Supplier 4 | 6 | 6 |
| Supplier 5 | 9 | 10 |
| Supplier 6 | 11 | 12 |
| Supplier 7 | 14 | 14 |
| Supplier 8 | 10 | 11 |
| Supplier 9 | 8 | 8 |
| Supplier 10 | 13 | 13 |
| Supplier 11 | 4 | 3 |
| Supplier 12 | 5 | 5 |
| Supplier 13 | 1 | 1 |
| Supplier 14 | 2 | 2 |

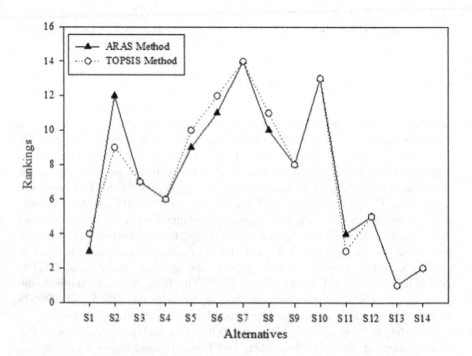

**FIGURE 10.2**   Ranking Results Obtained Using the ARAS and TOPSIS Methods.

data analysis went further. we are then able to understand the difference in both methods. It showed that there were considerable disparities in the outcomes and a high amount of uncertainty for other suppliers. Notwithstanding the fact that the obtained outputs explicitly make it apparent that there is a potential risk in adhering to the results of a single MCDM technique. Based on the results, both the ARAS and TOPSIS methods integrated with CRITIC are effective in solving supplier selection problems.

## REFERENCES

Abo-Sinna, Mahmoud A., and Azza H. Amer. 2005. "Extensions of TOPSIS for Multi-Objective Large-Scale Nonlinear Programming Problems." *Applied Mathematics and Computation* 162 (1): 243–256. doi:10.1016/j.amc.2003.12.087.

Boran, Fatih Emre, Serkan Genç, Mustafa Kurt, and Diyar Akay. 2009. "A Multi-Criteria Intuitionistic Fuzzy Group Decision Making for Supplier Selection with TOPSIS Method." *Expert Systems with Applications* 36 (8): 11363–11368. doi:10.1016/j.eswa.2 009.03.039.

Chatterjee, Prasenjit, and Shankar Chakraborty. 2013. "Gear Material Selection Using Complex Proportional Assessment and Additive Ratio Assessment-Based Approaches: A Comparative Study." *International Journal of Materials Science and Engineering* 1 (2): 104–111. doi:10.12720/ijmse.1.2.104-111.

Chen, Yuh Jen. 2011. "Structured Methodology for Supplier Selection and Evaluation in a Supply Chain." *Information Sciences* 181 (9): 1651–1670. doi:10.1016/j.ins.201 0.07.026.

Dadelo, Stanislav, Zenonas Turskis, Edmundas Kazimieras Zavadskas, and Ruta Dadeliene. 2012. "Multiple Criteria Assessment of Elite Security Personal on the Basis of ARAS and Expert Methods." *Economic Computation and Economic Cybernetics Studies and Research* 4: 65–87.

Dickson, Gary W. 1966. "An Analysis Of Vendor Selection Systems And Decisions." *Journal of Purchasing* 2 (1): 5–17. doi:10.1111/j.1745-493x.1966.tb00818.x.

Dogan, Ibrahim, and Nezir Aydin. 2011. "Combining Bayesian Networks and Total Cost of Ownership Method for Supplier Selection Analysis." *Computers and Industrial Engineering* 61 (4): 1072–1085. doi:10.1016/j.cie.2011.06.021.

Esbouei, Saber Khalili, and Abdolhamid Safaei Ghadikolaei. 2013. "An Integrated Approach Based on FAHP and ARAS Methods For." *ARPN Journal of Systems and Software* 3 (4): 53–56.

Ho, William, Xiaowei Xu, and Prasanta K. Dey. 2010. "Multi-Criteria Decision Making Approaches for Supplier Evaluation and Selection: A Literature Review." *European Journal of Operational Research* 202 (1): 16–24. doi:10.1016/j.ejor.2009.05.009.

Hwang, Ching-Lai, and Yoon, Kwangsun. 1981. *Methods and Applications a State-of-the-Art Survey.* Springer-Verlag, Berlin, Heidelberg. doi:10.1007/978-3-642-48318-9.

İç, Yusuf Tansel. 2012. "An Experimental Design Approach Using TOPSIS Method for the Selection of Computer-Integrated Manufacturing Technologies." *Robotics and Computer-Integrated Manufacturing* 28 (2): 245–256. doi:10.1016/j.rcim.2011.09.005.

Ighravwe, Desmond Eseoghene, and Moses Olubayo Babatunde. 2018. "Selection of a Mini-Grid Business Model for Developing Countries Using CRITIC-TOPSIS with Interval Type-2 Fuzzy Sets." *Decision Science Letters* 7 (4): 427–442. doi:10.5267/j.dsl.2018.1 .004.

Jayant, Arvind, and Mandeep Singh. 2015. "Use of Analytic Hierarchy Process (AHP) To Select Welding Process in High Pressure Vessel Manufacturing Environment." *International Journal of Applied Engineering Research* 10 (8): 5869–5884.

Jee, Dong Hyun, and Ki Ju Kang. 2000. "A Method for Optimal Material Selection Aided with Decision Making Theory." *Materials and Design* 21 (3): 199–206. doi:10.1016/s0261-3069(99)00066-7.

Kutut, V., E. K. Zavadskas, and M. Lazauskas. 2014. "Assessment of Priority Alternatives for Preservation of Historic Buildings Using Model Based on ARAS and AHP Methods." *Archives of Civil and Mechanical Engineering* 14 (2): 287–294. doi:10.101 6/j.acme.2013.10.007.

Mousavi-Nasab, Seyed Hadi, and Alireza Sotoudeh-Anvari. 2017. "A Comprehensive MCDM-Based Approach Using TOPSIS, COPRAS and DEA as an Auxiliary Tool for Material Selection Problems." *Materials and Design* 121: 237–253. doi:10.1016/j.matdes.2017.02.041.

Turskis, Zenonas, and Edmundas Kazimieras Zavadskas. 2010. "A New Fuzzy Additive Ratio Assessment Method (ARAS-F). Case Study: The Analysis of Fuzzy Multiple Criteria in Order to Select the Logistic Centers Location." *Transport* 25 (4): 423–432. doi:10.3846/transport.2010.52.

Weber, Charles A., John R. Current, and W. C. Benton. 1991. "Vendor Selection Criteria and Methods." *European Journal of Operational Research* 50 (1): 2–18. doi:10.1016/03 77-2217(91)90033-R.

# 11 Advances in Sheet Metal Stamping Technology: A Case of Design and Manufacturing of a Car Door Inner Panel Using a Tailor Welded Blank

*Tushar Y. Badgujar and Satish A. Bobade*

## CONTENTS

## 11.1   INTRODUCTION

The competitive and dynamic nature of today's automotive industry requires continuous development in products and operations. Due to government regulations and standards, there is an in-depth focus on reducing vehicle weight, improving its performance and lessening environmental issues (Kinsey, Liu, and Cao 2000). Automotive body components are generally manufactured by sheet metal stamping due to their characteristics such as high productivity, process and material utilisation efficiency and low cost of manufacturing. Significant reduction in weight can be achieved using more advanced material and alloys that are lighter in terms of weight. However, the cost of such high-strength materials is accordingly expensive, and these are not always the appropriate solution. The material cost in sheet metal forming contributes to approximately 80% of the final manufacturing cost (Kinsey and Wu 2011). Finally, crashworthiness and structural rigidity are the two performance parameters that are critical in achieving customer satisfaction and government regulations (Wang et al. 2018). There is a need to explore innovative solutions to attain low cost, high product quality and performance and weight reduction simultaneously. Using tailor welded blank could be one of these strategies. Not only automotive but also the aerospace and appliance manufacturing industries, among others, as well integrate tailor welded blanks into their design and manufacturing systems (Rooks 2001).

Two or more different sheets are welded together before final forming (Merklein et al. 2014). Blank tailoring means using different materials and composites for different sections of the same components. This ensures weight reduction along with provisioning extra reinforcement. There are many significant benefits of using tailor welded blanks (Zadpoor, Sinke, and Benedictus 2007):

- lighter and stronger components
- desirable functionality at a lower cost
- reduction in the number of parts
- increased part integrity

On the other hand, the disadvantage of TWB is the mixed nature of the blank where the thicker and thinner materials deform differently and tear prematurely at the time of stamping, which also results in weld line movement (Badgujar and Bobade, 2017).

In the manufacturing of TWBs, a blank is produced by joining several metal sheets with or without different thicknesses, materials and surface coatings (Babic et al. 2008). The welding of two sheets into one is the critical step in the preparation of TWB. The weld zone properties of the tailor welded blank play a vital role in the overall performance of the blank (Rojek et al. 2012; Cheng et al. 2007). Some welding methods used to create tailor welded blanks include laser electron beam welding (Chatterjee et al. 2009), mash seam welding (resistance welding) and friction stir welding (Zadpoor et al. 2008; Garware, Kridli, and Mallick 2010). Laser welding and mash seam are more popular methods that are used to weld blanks because of the low heat input applied in these methods. The low heat input of these welding processes results inless thermal distortion.

The joining (welding) is performed prior to forming of components. At the time of forming TWB, it is required to optimally distribute the material in order to achieve strength and cost optimisation. Ideally, the uniform distribution of material is significant (Badgujar and Wani 2018). Development with TWB means a lighter panel, high strength and welding before forming results in a reduction of production cost (Merklein et al. 2014; Dabhi, Thanki, and Patel 2014; Kinsey and Wu 2011). In the forming of the TWB, the formability of material plays a significant role. The formability of sheet metal depends on the thickness of the aforementioned sheet metal, its microstructure and external factors. Formability is the ease with which a given sheet of any material can be formed. The forming of sheet metal takes place due to plastic deformation achieved by mechanical means. A typical set-up consists of a platform and tooling is used where sheet metal is forced by a punch to get the shape of the die. The formability of sheet metal can be measured using several techniques (Gaied et al. 2009).

Some of the standardised sheet metal forming operations that are used to measure formability are i) stretch forming (Panda et al. 2007), e.g. hemispherical punch test and ii) deep drawing, (Patel, Shah, and Shah 2012), etc. Formability is also estimated by analytical methods such as i) the height of the form specimen in the Dome test (Lee et al. 2009), ii) the strain distribution across the deformed blank (i.e. forming behaviour), 3) the forming limit diagram (FLD) based on the strain measured, (Cikmis, Pepelnjak, and Hasanbegovic 2010; Ravi Kumar 2002), and more. In these methods, the sheet metal is required to deform until necking occurs or when the metal reaches its forming limit. All materials have a forming limit, which is defined as the maximum uniform strain adjacent to a localised neck or tear in a deformed specimen. The forming limit diagram method is a more common method used in the industry for measuring the formability of materials (Chan, Chan, and Lee 2003).

The FLD is used to determine critical areas of strain. The forming limit curve of the material is plotted in the FLD, which refers to the forming limit of the specific material under a full range of strain states. Higher forming limit curve of the sheet metal results in better formability. The forming limit curve shape depends on the material properties, sheet thickness and size, the strain path and strain gradients. FLD is used in formability analysis to be able to diagnose problems by comparing the failure-prone areas in the FLD. This approach is used to determine the severity of the potential problems due to factors such as lubrication, tooling, material properties and thickness. The FLD illustrates that the strain path and the strain at localised necking are a function of the minor strain. Moreover, some important points related to an FLD are the modes of stretching. The right side of an FLD shows strain in both the minor and major directions of tension. On the left side of the FLD, the minor strain is negative and similar to the uniaxial tensile condition. Finally, the lowest point in FLC occurs under the plane strain stretching condition where the major strain is the lowest, and the minor strain remains unchanged at zero.

## 11.2  RESEARCH OBJECTIVE

The objective of the present research work is to propose the design of TWB for a car door inner panel using FEA. The stress distribution in a panel manages to reduce the weight of the door without compromising strength and the subsequent manufacturing of the panel.

## 11.3  METHODOLOGY

The critical steps followed in the design and manufacturing of the inner door panel are as below:

1. Identification of panel specifications
2. Feasibility checking of specifications and appropriate modifications if required in the specifications in order to improve usefulness
3. Simulation of stamping process using AutoForm software
4. Die designing
5. Production of the panel, if the deviation from the specification is observed, then restart the cycle and perform the third step with the modified specification.
6. Inspecting produced panel for quality checks

## 11.4  FINITE ELEMENT ANALYSIS

In the present study, software called AutoForm is used to conduct the analysis. The materials chosen for this panel are of two thicknesses, 1.4 mm and 0.7 mm. The thicker materials are used nearer the door hinge area (Figure 11.1), as this area experiences maximum working stress, while other areas are used for accessories mounting. The line of TWB is selected and provided to the simulation as input.

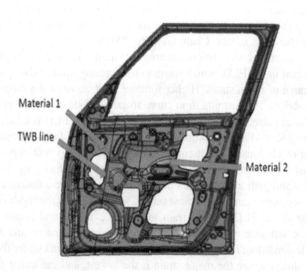

**FIGURE 11.1**  Door Panel.

FEM analysis starts with the identification of material properties. Extra deep drawn (EDD) steel is used as a blank material.

EDD steel contains a very low carbon content and is chemically stable, possesses exceptional formability, excellently homogeneous and strong against fatigue and denting. The blanks made by EDD are quite suitable for deep drawing, because the resistance to thinning remains high during this process (Dabhi, Thanki, and Patel 2014; Ma and Guan 2016).

Some important information with regard to material properties is as follows:

| | |
|---|---|
| Material Name | : MM 22 EDD |
| Material Thickness | : 1.4 mm and 0.7 mm |
| Maximum Yield Stress | : 210 MPa |
| Maximum Tensile Strength | : 350 Mpa |
| Maximum Hardness | : 50 HRC |
| Poison's Ratio | : 0.33 |
| % Carbon | : 0.08 |
| % Manganese | : 0.4 |
| % Sulphur | : 0.03 |
| % Phosphorus | : 0.06 |

### 11.4.1 FORMABILITY CHECKING FOR 1.4 MM-THICK SECTION OF A DOOR PANEL

Formability checking of the panel was done in two steps, as the effect of forming is required to be studied in two areas. The first section of the panel with 1.4 mm sheet thickness is analysed, and Figure 11.2 shows the formability limit diagram (FLD)

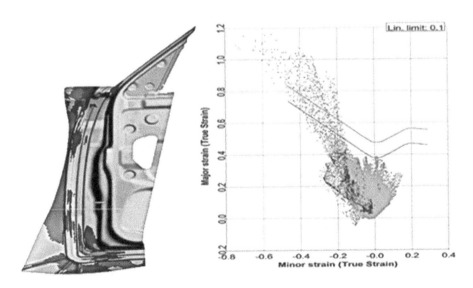

**FIGURE 11.2** Formability Checking for 1.4 mm-Thick Material.

for 1.4 mm material thickness. It is observed that most of the area falls under a safe zone. The thickening is outside of the functional area of a panel. There are certain places identified where a high chance of metal compression is recognized. The major strain is below 0.5, which is within the acceptable limit. Moreover, the results show that 1.4 mm material does not undergo any splits. Excessive thinning is within 0.2 mm, which is adequate, and the risk of splits is within the limit.

### 11.4.2 FORMABILITY CHECKING FOR 0.7 MM-THICK SECTION OF A DOOR PANEL

Figure 11.3 shows the formability limit diagram (FLD) for 0.7 mm thickness. A majority of forming lies in the safe zone, except for tearing in a scrap zone (i.e. the window region) which is intentional. The window region material is removed in subsequent steps. The stress distribution is fairly uniform, and no stress concentration observed.

### 11.4.3 COMBINED FORMABILITY CHECKING

When formability checking of the complete door panel is done, it differs from the individual material analysis. It shows that most of the area falls under the safe zone, but the thickening area has increased outside the functional area. There are certain places where a high chance of metal compression is identified, which is the same as individual material. The major strain is below 0.5, which is within the acceptable limit. The material splitting is in the scrap zone, which likewise is predictable.

### 11.4.4 TWB LINE MOVEMENT

Weld line movement during forming is critical for a TWB. This analysis shows a TWB line flow in draw operation. The weld line movement data is an input for die

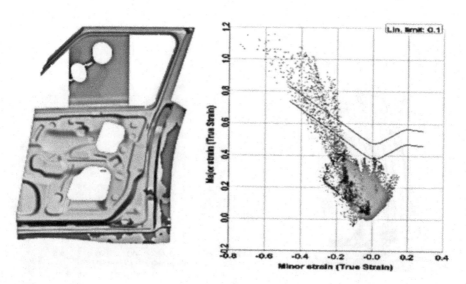

**FIGURE 11.3**   Formability Checking for 0.7 mm-Thick Material.

design so that corresponding modifications can be made in die designing. As per line movement, a corresponding relief is done in subsequent dies to avoid any fouling. In the TWB line movement, the maximum movement observed is 20.75 mm, which is more than the generally permitted movement of 5.0 mm. If it moves beyond the permissible limit, then it might get foul in a subsequent operation and affect forming of other regions or damage the dies. If fouling is accepted at this stage, then relief is required in the next step. In the present case, movement is observed within the allowable limit at two places. However, in the third place, it is moving at approximately 20.0 mm. Because of this, all stages in the upper pad are relieved in order to avoid any damage. It is moving due to scrap holes provided, which experiences shearing (Figure 11.4).

## 11.4.5 Thinning of TWB in Forming

Thinning is a process that causes a reduction in material thickness at a particular point. Generally, 20% of material thickness thinning is permitted in the forming of the TWB. If thinning of more than 20% occurs, then it can lead to material cracking and tearing defects. In the present simulation, it is measured as 0.2 mm thinning to 0.05 mm thickening. The maximum thinning is observed in the window area due to excess allowed material flow. Thickening is observed in the outside panel area, which is not at all relevant. As for the corner area on the window, the thinning overserved, but this is rectified before the final simulation.

## 11.4.6 Springback Effect

Springback is one of the most critical phenomenon observed in forming operations. It is not easy to predict the exact value of springback, which could occur in a

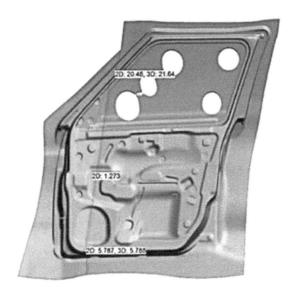

**FIGURE 11.4** Weld Line Movement.

particular operation. In the present case, the springback observed is of a twisted type. One end showed -4.0 mm, whereas another end measured +5.56 mm. It was subsequently decided to take corrective action at the time of actual panel trials.

From the various analyses above, it is concluded that both 1.4 mm and 0.7 mm thickness materials showed adequate formability without any failure with given parameter values in both individual and combined formability checking. No major defect or fault is observed in time steps, and excellent forming of a panel is observed. The TWB weld line movement is around 5 mm, except in the window area, which measures approximately 20 mm. The movement of the weld line in this location is due to intentional scrap holes provided in the window area. Moreover, the window area identified tearing as a result of scrap holes. Twisted springback was found in the window area and no action was taken for its elimination. It has been decided to take action regarding this matter after the actual panel analysis.

## 11.5   DIE DESIGN

After a successful simulation, die designing is accomplished. The simulation is an input for die design. The first step in die design is dies layout preparation, where the sequence of operation in each die is determined. In this step, the panel flow is also selected. This is one of the most critical stages in die designing. Every operation is critically reviewed so that it is feasible to work while considering the type of press, kind of operation (manual or automated) and location of production, among other factors. At this point, the progression of the operation is then decided. This panel has vertical trimming, vertical pierce with cam pierce, a flange and restrike operation.

It would not be possible to perform all of the operations at a single time; hence, the steps are broken down. As per simulation data, the operation will be completed in the five stages mentioned below:

| | |
|---|---|
| TWB Blank Preparation | : First Operation |
| Draw | : Second Operation |
| Trim and Pierce | : Third Operation |
| Cam Pierce | : Fourth Operation |
| Flange, Restrike and Cam Pierce | : Fifth Operation |

### 11.5.1   DRAW DIE DESIGN

Draw die design is the first operation performed on the TWB, while the blank is prepared as specified in advance by using laser welding. This draw design is different from the general draw design. Elementwise, it is similar to a simple draw design. It has three main elements, namely the lower punch, blank holder and upper cavity. All of these elements have a basic casting grade structure as FG300. However, the blank holder and upper cavity each have 14 SKD-11 alloy steel material inserts fitted in the area where the blank thickness is 1.4 mm. This is intentionally accomplished considering the following reasons:

- For sheet thickness above 1.2 mm, steel inserts are recommended.
- There is a possibility of casting wear due to material flow.
- There is a possibility of generating defects like wrinkle and crack formation due to insufficient material holding.
- The smooth working of the die must be maintained throughout its life span.

### 11.5.2 FLANGE-RESTRIKE DIE DESIGN

Die design combines left-hand (LH) and right-hand (RH) side panels in a single die. The lower die has the flange restrike inserts along with die buttons fitted in it for cam pierce. Flanging takes place against the lower and upper inserts. The upper die consists of a pad and flange inserts, and cam pierce punches are mounted on a standard cam slider which operates by drivers fitted on the lower die. The pad touches and holds lower inserts. After the flange restrike is done, it is succeeded by cam pierce operation.

## 11.6 MANUFACTURING

After achieving satisfactory results of the simulation, the abovementioned parameters are taken for actual panel production. Trials are planned on the hydraulic press, as it is possible to make an observation at particular trials, which is not achievable with mechanical presses. Virtual trial results of the simulation are verified in actual tests at the same intervals. The press used for manufacturing is the Hindustan Machine Tool hydraulic press of 1300 Ton as pictured in Figure 11.5.

The following are the vital parameters provided by the simulation for panel production:

**FIGURE 11.5**  Hydraulic Press.

| Approximate Forming Pressure | : 665 T (Simulation Reaction) |
|---|---|
| Approximate Blank Size (mm) | : 440 × 930 × 0.7 and 300 × 400 × 1.4 |
| Binder Stroke | : 120 mm |
| Material Properties | : As per die layout |

## 11.6.1 Draw Panel Production

Draw die in any panel consumes maximum lead time. Die face working is the most critical and time-consuming part of the work. Draw panel uses the TWB. Die has a total number of 66 cushion pins, and it is loaded on the press. After die face working, it is observed to have taken the exact time at the steps on the press as that likewise taken in simulation to check defect generation as well as panel formation at various heights. The following defects are developed in the door panel.

## 11.6.2 Springback in the Window Area

Springback occurs during sheet metal forming. This is the phenomenon in which the formed sheet metal attempts to go back to its original position upon the removal of forces (Gan and Wagoner 2004; Panthi et al. 2010). It affects the dimensional accuracy of the final parts. Springback can be approximated based on blank-holder force, mould parameters and material properties (Padmanabhan et al. 2009). At the start of the forming process, preventive measures can be taken. The car door panel manufacturing also encounters springback. Certain actual problems with regards to springback are the following:

- difficulty in the prediction of the final shape and dimension of the panel, and
- challenges in designing the appropriate tools to compensate for springback

The springback defect is observed in the window area due to less strength in that location, because scrap holes are present. The lower corner where the window area starts is a weaker zone in forming as a limited holding. The area above the line in Figure 11.6 was discovered to measure up by 0.8 mm, as compared with panel data inspection. The other end went down by 2.0 mm. Simulation values were 5.56 mm and -4 mm for respective places.

## 11.6.3 Wrinkles and Crack in the Formed Panel

The wrinkle and cracks are inversely proportional to each other. If the blank holding is reduced, then wrinkles will generate, and if the holding is increased, then the result is crack formation. Therefore, complete elimination of wrinkles and cracks at the simulation stage is not possible (Dabhi, Thanki, and Patel 2014; Chen and Liao 2000). However, after simulation correction, this occurs in panel production, and its removal requires trial and error iteration in the actual panel production with the help of design and simulation techniques. As shown in Figure 11.7, wrinkles are observed in 'a' to 'd', whereas cracks found in 'e'.

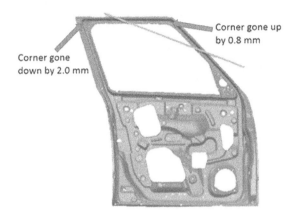

**FIGURE 11.6**   Springback Defect in the Window Area.

## 11.6.4  DEFECTS RECTIFICATION

For the rectification of defects, setting parameters are changed to new specifications as listed below:

| | |
|---|---|
| Approximate Forming Pressure | : 670 T |
| Approximate Blank Size | : No change |
| Binder Stroke | : 122 mm |
| Material Properties | : As per die layout |

Along with these altered parameters, dies are modified in order to eliminate defects as discussed below.

## 11.6.5  SPRINGBACK RECTIFICATION

To reduce springback, the die face is displaced in the opposite direction of springback, with the same value. A multiplication factor defines the actual compensation. In this case, negative forming must be executed. Welding and machining of the upper tool and lower punch material are removed by machining. In

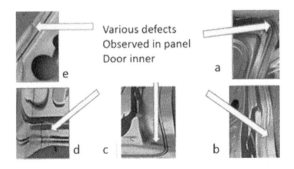

**FIGURE 11.7**   Wrinkle and Crack Defects Observed in Panel.

Springback

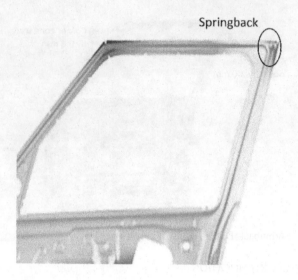

**FIGURE 11.8**   Springback Before Correction.

Figure 11.8, the highlighted area presented a 2.462 mm springback in the simulation and -0.582 mm at the other corner. Springback is eliminated after implementing correction. Figures 11.8 and 11.9 show springback before and after rectification (Wang et al. 2016).

## 11.6.6   Wrinkle and Crack Rectification

In the experimental production of a panel, it is observed that improper pad holding played a vital role in defect generations. After several trial and error iterations on

**FIGURE 11.9**   Springback After Correction.

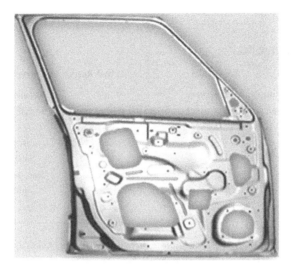

**FIGURE 11.10**   Manufactured Door Panel.

the press, the finalised area requires holding improvement. For increased holding, s 0.5 mm surface is created to attain hard-hitting. For the same purpose, the upper tool is welded and machined.

## 11.7   VALIDATION

The corrective steps are taken and validated by actual panel production. The findings are as follows:

Panel springback is corrected by welding and machining in the affected zone, and results are consistent as per expectation. Figure 11.10 shows the produced door panel.

1. Wrinkles and cracks are eliminated by hard-hitting.
2. Panel thinning discovered within 20% of panel thickness is considered acceptable.
3. The panel produced on the press is free from any other defects, and at no part is geometry deviation observed.
4. This produced panel is inspected on a coordinate measuring machine (CMM), and the report shows that it is within the acceptable limit.

## 11.8   RESULTS AND VALIDATION

There is a significant similarity found among the results that were obtained by the actual panel produced and FEA analysis. The comparison is shown in Table 11.1. It was observed that FEA results for the TWB weld line shifting and material thinning are complementary with each other. However, prediction failed in case of defect generation, such as springback and cracks. Those developed defects are studied using simulation, and changes are proposed. After the implementation of the

**TABLE 11.1**

**Comparison of Results**

| Sr. No | Description | FEA Results | Actual Results | Corrective action |
|---|---|---|---|---|
| 1 | TWB line shifting | 5.0 mm max | 4.8 mm | no action needed |
| 2 | material thinning | 20% of panel thickness | 15% of panel thickness | no action needed |
| 3 | springback | not observed | present in some areas | as discussed in Section 11.6.5 |
| 4 | cracks | bead radius maintained at 3 mm and height retained at 5 mm | cracks observed | bead radius changed to 6 mm and height reduced to 2 mm. |

aforementioned changes, improvement is observed and leads to the complete elimination of defects and the production of a defect-free panel.

## 11.9 CONCLUSION

This chapter presented an FEA-assisted design and the manufacturing of a tailor welded blank tooling. It proved that FEA simulation is a useful tool for virtual formability checks of a selected material. Time steps are also efficient in predicting the defects generated at various time steps in virtual tryouts. The TWB weld line is movement anticipated, and compensation is a critical phenomenon during the simulation process. The actual panel produced developed multiple defects such as springback, wrinkles and cracks. Suggested measures successfully eliminated these defects. The production of an actual panel validated the simulation results. The same methodology is beneficial in developing a generalised approach for various industrial blank-related problems, where the use of one or more material/s with similar or different sheet thickness can optimise parameters, such as strength, weight and productivity. Finally, it must be noted that working with springback and weld line shifting requires proper care.

## REFERENCES

Babic, Z., S. Aleksandrovi, M. Stefanovi, and M. Sljivic. 2008. "Determination of Tailor Welded Blanks Formability Characteristics." *Journal for Technology of Plasticity* 33 (1).

Badgujar, Tushar Y., and Satish A. Bobade. 2017. "Tailor Welded Blanks (TWBs) for a Sheet Metal Industry an Overview." *International Journal of Advance Research and Innovative Ideas in Education* 3 (3): 3771–3779.

Badgujar, Tushar Y., and Vijay P. Wani. 2018. "Stamping Process Parameter Optimization with Multiple Regression Analysis Approach."*Materials Today: Proceedings* 5 (2): 4498–4507. https://doi.org/10.1016/j.matpr.2017.12.019.

Chan, S. M., L. C. Chan, and T. C. Lee. 2003. "Tailor-Welded Blanks of Different Thickness Ratios Effects on Forming Limit Diagrams." *Journal of Materials Processing Technology* 132 (1–3): 95–101. https://doi.org/10.1016/S0924-0136(02)00407-7.

Chatterjee, Sujit, Rajib Saha, M. Shome, and R. K. Ray. 2009. "Evaluation of Formability and Mechanical Behavior of Laser-Welded Tailored Blanks Made of Interstitial-Free and Dual-Phase Steels." *Metallurgical and Materials Transactions A: Physical Metallurgy and Materials Science* 40: 1142–1152. https://doi.org/10.1007/s11661-009-9808-2.

Chen, Fuh Kuo, and Yeu Ching Liao. 2000. "Three-Dimensional Finite Element Analysis of Wrinkling in a Stamping Process." SAE Technical Paper 2000-01-0777 : 1–12. https://doi.org/10.4271/2000-01-0777.

Cheng, C. H., M. Jie, L. C. Chan, and C. L. Chow. 2007. "True Stress-Strain Analysis on Weldment of Heterogeneous Tailor-Welded Blanks-a Novel Approach for Forming Simulation." *International Journal of Mechanical Sciences* 49 (2): 217–229. https://doi.org/10.1016/j.ijmecsci.2006.08.012.

Cikmis, Amra Talic, Tomaz Pepelnjak, and Suad Hasanbegovic. 2010. "Experimental Determination of Forming Limit Diagram." 14th International Research/Expert Conference, Mediterranean Cruise.

Dabhi, Hareshkumar O., Shashank J. Thanki, and Vimal K. Patel. 2014. "Experimental and Numerical Investigation of Tailor Welded Blanks [TWBs]" *International Journal of Advanced Mechanical Engineering* 4 (6): 589–599.

Gaied, Sadok, Jean Marc Roelandt, Fabrice Pinard, Francis Schmit, and Mikhael Balabane. 2009. "Experimental and Numerical Assessment of Tailor-Welded Blanks Formability." *Journal of Materials Processing Technology* 209 (1): 387–395. https://doi.org/10.1016/j.jmatprotec.2008.02.031.

Gan, Wei, and R. H. Wagoner. 2004. "Die Design Method for Sheet Springback." *International Journal of Mechanical Sciences* 46 (7): 1097–1113. https://doi.org/10.1016/j.ijmecsci.2004.06.006.

Garware, M., G. T. Kridli, and P. K. Mallick. 2010. "Tensile and Fatigue Behavior of Friction-Stir Welded Tailor-Welded Blank of Aluminum Alloy 5754." *Journal of Materials Engineering and Performance* 19: 1161–1171. https://doi.org/10.1007/s11665-009-9589-1.

Kinsey, Brad, Zhihong Liu, and Jian Cao. 2000. "Novel Forming Technology for Tailor-Welded Blanks." *Journal of Materials Processing Technology* 99 (1–3): 145–153. https://doi.org/10.1016/S0924-0136(99)00412-4.

Kinsey, Brad L., and Xin Wu. 2011. *Tailor Welded Blanks for Advanced Manufacturing. Tailor Welded Blanks for Advanced Manufacturing.* United Kingdom: Woodhead Publishing Limited. https://doi.org/10.1533/9780857093851

Lee, Wonoh, Kyung Hwan Chung, Daeyong Kim, Junehyung Kim, Chongmin Kim, Kazutaka Okamoto, R. H. Wagoner, and Kwansoo Chung. 2009. "Experimental and Numerical Study on Formability of Friction Stir Welded TWB Sheets Based on Hemispherical Dome Stretch Tests."*International Journal of Plasticity* 25 (9): 1626–1654. https://doi.org/10.1016/j.ijplas.2008.08.005.

Ma, Xiang Dong, and Ying Ping Guan. 2016. "Theoretical Prediction and Experimental Investigation on Formability of Tailor-Welded Blanks." *Transactions of Nonferrous Metals Society of China (English Edition)* 26 (1): 228–236. https://doi.org/10.1016/S1003-6326(16)64108-0.

Merklein, Marion, Maren Johannes, Michael Lechner and Andreas Kuppert. 2014. "A Review on Tailored Blanks - Production, Applications and Evaluation." *Journal of Materials Processing Technology* 214 (2): 151–164. https://doi.org/10.1016/j.jmatprotec.2013.08.015.

Padmanabhan, R., M. C. Oliveira, H. Laurent, J. L. Alves, and L. F. Menezes. 2009. "Study on Springback in Deep Drawn Tailor Welded Blanks." *International Journal of Material Forming* 2: 829. https://doi.org/10.1007/s12289-009-0566-x.

Panda, Sushanta Kumar, D. Ravi Kumar, Harish Kumar, and A. K. Nath. 2007. "Characterisation of Tensile Properties of Tailor Welded IF Steel Sheets and Their Formability in Stretch Forming." *Journal of Materials Processing Technology* 183 (2–3): 321–332. https://doi.org/10.1016/j.jmatprotec.2006.10.035.

Panthi, S. K., N. Ramakrishnan, Meraj Ahmed, Shambhavi S. Singh, and M. D. Goel. 2010. "Finite Element Analysis of Sheet Metal Bending Process to Predict the Springback." *Materials and Design* 31 (2): 657–662. https://doi.org/10.1016/j.matdes.2009.08.022.

Patel, B. C., Jay Shah, and Harindra Shah. 2012. "Review on Formability of Tailor-Welded Blanks." *International Journal on Theoretical and Applied Research in Mechanical Engineering (IJTARME)* 1: 89–94.

Ravi Kumar, D. 2002. "Formability Analysis of Extra-Deep Drawing Steel." *Journal of Materials Processing Technology* 130–131: 31–41. https://doi.org/10.1016/S0924-013 6(02)00789-6.

Rojek, J., M. Hyrcza-Michalska, A. Bokota, and W. Piekarska. 2012. "Determination of Mechanical Properties of the Weld Zone in Tailor-Welded Blanks." *Archives of Civil and Mechanical Engineering* 12 (2): 156–162. https://doi.org/10.1016/j.acme.2012.04 .004.

Rooks, Brian. 2001. "Tailor-Welded Blanks Bring Multiple Benefits to Car Design." *Assembly Automation* 21 (4): 323–329. https://doi.org/10.1108/EUM0000000006014.

Wang, Hangyan, Hui Xie, Wei Cheng, Qiming Liu, and Yunfei Shen. 2018. "Multi-Objective Optimisation on Crashworthiness of Front Longitudinal Beam (FLB) Coupled with Sheet Metal Stamping Process." *Thin-Walled Structures* 132 (April): 36–47. https://doi.org/10.1016/j.tws.2018.07.050.

Wang, H., J. Zhou, T. S. Zhao, L. Z. Liu, and Q. Liang. 2016. "Multiple-Iteration Springback Compensation of Tailor Welded Blanks during Stamping Forming Process." *Materials and Design* 102: 247–254. https://doi.org/10.1016/j.matdes.2016.04.032.

Zadpoor, A. A., J. Sinke, and R. Benedictus. 2007. "Mechanics of Tailor Welded Blanks: An Overview." *Key Engineering Materials* 344: 373–382. https://doi.org/10.4028/ www.scientific.net/KEM.344.373.

Zadpoor, Amir Abbas, Jos Sinke, Rinze Benedictus, and Raph Pieters. 2008. "Mechanical Properties and Microstructure of Friction Stir Welded Tailor-Made Blanks." *Materials Science and Engineering A* 494 (1–2): 281–290. https://doi.org/10.1016/j.msea.2 008.04.042.

# 12 Advanced Fabrication of Banana Fibre-Based Hybrid Composites

*Upendra Sharan Gupta, Mohit Dhamarikar,*
*Amit Dharkar, Siddhartha Chaturvedi,*
*Sudhir Tiwari, and Rajeev Namdeo*

## CONTENTS

## 12.1 INTRODUCTION

Natural fibres are fibres derived from either plants or animals (Bhattacharyya, Subasinghe, and Kim 2015). Majorly composed of cellulose, hemicellulose and lignin plant fibres have earned themselves the name lignocellulosic fibres (Satyanarayana et al. 1990; Jones et al. 2017). Natural fibres have gained the attention of industrial and academic researchers due to growing concerns regarding the environment, ecology and the economy. Consequently, costly non-biodegradable and non-renewable materials are being replaced with sustainable, natural eco-friendly and cost-friendly materials, such as fibres (Varghese and Mittal 2017). 'Natural fibres' are favoured over 'synthetic fibres' due to adequate specific modulus values, low specific weight, considerably high toughness values of plant fibres along with abundance in availability and cost-friendliness (Kilinç, Durmuşkahya, and Seydibeyoğlu 2017). Using natural fibre-based composites to manufacture interior and exterior parts can reduce the overall weight of the vehicle by up to 40% (Huda et al. 2008). Construction, furniture, packaging and automotive industries are experimenting with various ways to utilise natural fibre biocomposites over the conventionally used materials (Kim 2012).

Natural fibre composites present considerable environmental advantages compared to synthetic fibres due to biodegradability. Because of their economic friendliness, low specific weight and high specific strengths, they possess the potential to be used for manufacturing the interior part of aircraft (Balakrishnan et al. 2016).

Banana belongs to the Musaceae family and has around 300 species existing; out of which, only 20 varieties are used for consumption (Ramesh 2018). Banana fibre is a resource developed from bio-waste, and can thus be obtained without any additional cost input for industrial applications (Joseph et al. 2002). Kulkarni et al. (1983) conducted an experimental investigation on banana fibre and discovered that the elastic modulus of the banana fibre exists in the range of 27 to 32 $GN/m^2$, while the ultimate tensile strength lies in the range of 711 to 789 $MN/m^2$, and percentage elongation range is 2.5 to 3.7% for fibres of the diameter 50 to 250 μm. Jústiz-Smith, Virgo, and Buchanan (2008) conducted various fibre characterisation tests and investigated that banana fibre had ash content of 8.3% by weight and carbon content of 50.9% by weight, which are higher amounts than those present in coconut and bagasse fibres. Percentage water uptake was also discovered to be the least in banana fibre.

The motive of hybridisation is to fabricate a new material out of two constituents that will consequentially attain the same properties as those of its constituents while simultaneously overcoming their limitations (Kretsis 1987). Kureemun et al. (2018) studied the effects of hybridising flax fibre over carbon fibre and noted that an improvement of up to 50% in the strength and stiffness for natural flax fibre was possible with hybridisation. The hybridisation of natural fibres is an effective means of tailoring material properties according to service requirements (Fu et al. 2001). The hybridisation of natural fibre not only improves the strength of the banana fibre, but has also reduced its cost and transforms it into an eco-friendly composite (Navaneethakrishnan, Selvam, and Jaisingh 2015; Madhukiran, Rao, and Madhusudan 2013; Idicula et al. 2005). Natural fibre reinforced hybrid composite garnered the attention of the automotive and aerospace sectors as a further step in decreasing the weight and increasing the fuel efficiency of the vehicle and revolutionising its design as eco-friendly (Panthapulakkal et al. 2017).

The method of fabrication plays an important role in increasing the efficacy of the hybridisation process. Various methods of fibre composite fabrication are commercially known and are industrially mainly used for the fabrication of synthetic composites. It is observed that the hand layup method and compression moulding are the most commonly used procedures for the fabrication of banana fibre hybrid composites due to their simplicity, out of which compression moulding is significantly preferred commercially due to control over curing time by managing temperature and pressure applied (Jamir, Majid, and Khasri 2018). However, other methods of composite preparation are not usually used by researchers in preparing banana fibre hybrid composites. This chapter aims to discuss the diverse effects of hybridising among other natural, as well as synthetic, fibres with banana fibre. Moreover, various composite fabrication methods are discussed to identify the most ideal process in terms of geometric accuracy in order to establish hybridisation as an efficient process in enhancing the mechanical characteristics of industrially viable banana fibre from a commercial standpoint.

## 12.2 MODERN FABRICATION OF COMPOSITES

Banana fibre hybrid composites can be prepared by various methods which are briefly discussed below along with each procedure's dimensional advantages of fabrication. Even though the method of fabrication does not appear to have any significant effects on the mechanical properties of the product, the properties are indeed enhanced due to the hybrid effect alone, as discussed in Section 12.3. However, the method of fabrication used highly affects the degree of dimensional accuracy and aesthetic characteristics of the finished product. Thus, the methods of banana fibre hybrid composite fabrication are discussed in detail.

### 12.2.1 HAND LAYUP METHOD

Banana fibre hybrid composites are usually prepared by the hand layup method. The hand layup method is one of the oldest yet one of the simplest methods used in manufacturing banana fibre composites (Jamir, Majid, and Khasri 2018). In this process, each layer is laid on the other by hand (Elkington et al. 2015). Although straightforward and reliable, the hand layup process requires more labour and takes a comparatively longer time than the advanced manufacturing techniques. The rate of production depends on the skill of the worker as well as the complexity of the desired structure (Van Hattum, Regel, and Labordus et al. 2011). In this method, the resin is firstly coated over the mould surface. After which, the reinforcing layer of banana fibre is laid by hands and distributed uniformly in the mould. Next, the binding resin is added to the mould. The air entrapped within the layers is then degassed by the means of squeegees or rollers. The resin, which cures at room temperature, is usually adopted in this method. The catalyst aids the initiation of curing of the composite, thus hardening the composite without the application of 'external heat' (Balasubramanian, Sultan, and Rajeswari 2018). Before mixing the matrix and the banana fibre, the density of the obtained composite should be calculated so that the optimum amount of raw material is used. The hand layup method is a fitting procedure for fabricating banana fibre hybrid composites incorporating both thermosetting and thermoplastic resins (Asim et al. 2017) whose schematic representation is shown in Figure 12.1.

### 12.2.2 COMPRESSION MOULDING

Compression moulding is the most common method used in manufacturing large components. Compression moulding can be categorised as hot and cold moulding. In the cold moulding technique, only pressure is applied; whereas the hot moulding technique involves the application of both pressure and heat (Asim et al. 2017). Composites with a thickness of 1 mm to 10 mm can be fabricated by this method. The 'curing temperature' for the 'cold and hot moulding process' usually lies in the range of 40°C to 50°C and 80°C to 100°C, respectively and the curing time is approximately 1–2 hours (Jamir, Majid, and Khasri 2018). Compression moulding is generally used to manufacture banana fibre-reinforced polymer matrix composite due to its 'high reproducibility' and 'low cycle time' and is even more well-known

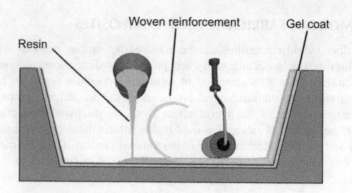

**FIGURE 12.1** Schematic Representation of the Hand Layup Method. (Reprinted from *Durability and Life Prediction in Biocomposites, Fibre-Reinforced Composites and Hybrid Composites, 1st edition,* Raji, Marya, Hind Abdellaoui, Hamid Essabir, Charles Amani Kakou, Rachid Bouhfid and Abou El Kacem Qaiss, *Prediction of the Cyclic Durability of Woven-Hybrid Composites.*, 27-62, 2018, with permission from Elsevier)

for its 'low fibre attrition' and 'speed'. Furthermore, compression moulded composites have high impact strength (Ismail et al. 2015). Figure 12.2 presents the schematic of the machine used for the fabrication of composites.

## 12.2.3 RESIN TRANSFER MOULDING

Resin transfer moulding is a slight variation of compression moulding, in which the mould is closed in advance, and the liquid thermoset resin is transferred into the closed

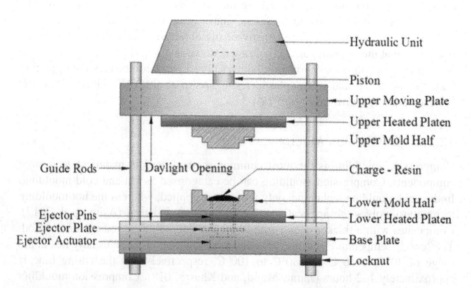

**FIGURE 12.2** Compression Moulding Setup/Machine. (Reprinted from Compression Moulding. Applied Plastics Engineering Handbook, first edition, Tatara, 2011. With permission from Elsevier)

mould (Dai and Fan 2013; Tanzi, Farè, and Candiani 2019). It is a low-pressure and low-temperature method (Dai and Fan 2013). Resin transfer moulding is a medium pressure process and works in the range of 3.5–7 bar. This method is preferred for manufacturing complex 3D parts (Erden and Ho 2017). Banana fibre composites with favourable mechanical properties and attractive finish can be manufactured using the RTM processing technique, which is illustrated in Figure 12.3 (Oksman 2001).

## 12.2.4 INJECTION MOULDING

Injection moulding is a rapid process that operates at low pressure. Similar to resin transfer moulding, it is also a closed mould process (Samivel and Babu 2013). Injection moulding is a prevailing method used for manufacturing banana fibre-plastic composite parts. Injection moulded banana fibre reinforced composites have exceptionally high dimensional accuracy (Liu 2012). As illustrated in Figure 12.4, in this process, the 'plastic material' is first heated until it turns into a 'viscous melt'. The liquid is then forced into a 'closed mould cavity' resembling the profile of the desired product. The material is allowed to cool there until it solidifies. Finally, the finished solid composite is then extracted from the mould (Ebnesajjad and Ebnesajjad 2003). Injection moulded products have an edge over compression moulded components in terms of accuracy, surface finish, less warpage and minimal shrinking (Faruk et al. 2012).

## 12.2.5 VACUUM BAG MOULDING

Vacuum bag moulding is a process of laminated composite manufacturing that is especially undertaken in manufacturing industries. A partially cured thermosetting

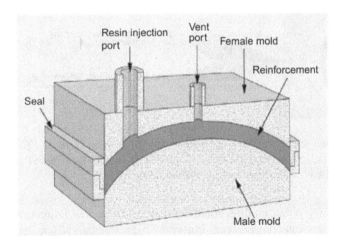

**FIGURE 12.3** Resin Transfer Moulding. (Reprinted from Fibre Technology for Fibre-Reinforced Composites, First Edition, Erden, Seçkin and Kingsley Ho, Fibre Reinforced Composites, 51-79, 2017, With permission from Elsevier)

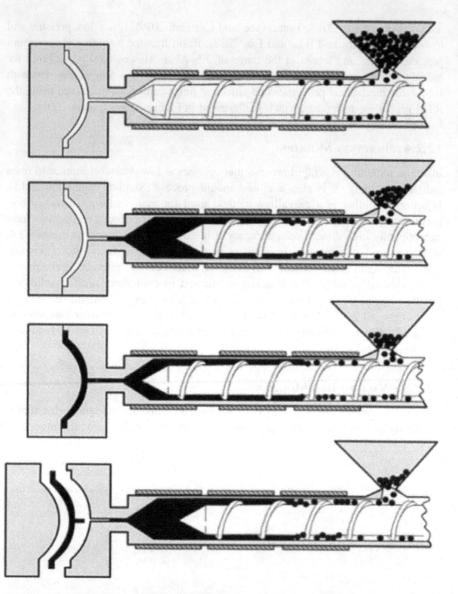

**FIGURE 12.4** Injection Moulding. (Reprinted from Manufacturing Techniques for Polymer Matrix Composites, First Edition, Liu, S.-J, Injection Moulding in Polymer Matrix Composites, 15-46, 2012, with kind permission from Elsevier)

polymer bearing either unidirectional continuous fibres or a bidirectional fabric is used as the raw material prepreg. Figure 12.5 presents a vacuum bag moulding setup. In this process, plies are cut from the rolls of prepreg and subsequently stacked. Thereafter, these are covered by a vacuum bag which is in the form of a polymer film. The enclosed material in the vacuum bag is cured in an oven or

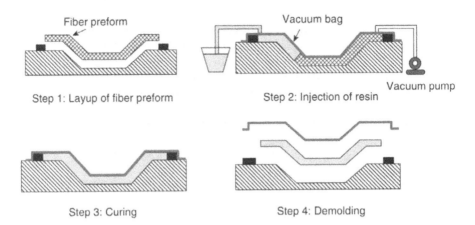

**FIGURE 12.5** Vacuum Bag Moulding. (Reprinted from Developments in Fibre-Reinforced Polymer (FRP) Composites for Civil Engineering, First Edition, El-Hajjar, R., H. Tan, Advanced Processing Techniques for Composite Materials for Structural Applications, 54-76, 2013, with kind permission from Elsevier)

autoclave where a controlled atmosphere is achieved to obtain the final product as a solid laminate (Mallick 2010). The performance parameters of accuracy of vacuum bag moulded components include the position of the injection tube and air outlet, permeability of fibre, viscosity of resin, etc. (El-Hajjar, Tan, and Pillai 2013). Banana fibre based hybrid composites are also manufactured using this technique.

## 12.3  REVIEW OF PAST WORK

A group of researchers obtained the composite by hybridising jute fibre, with a modulus of 55 GPa and a density of 1,300 Kg/m$^3$, with banana fibre using an epoxy resin of modulus 3.42 GPa and a density of 1,100 Kg/m$^3$ as reinforcement in 0:100, 25:75, 50:50, 75:25 and 100:0 weight ratios (Boopalan, Niranjanaa, and Umapathy 2013). The slab of the composite formed was successively used for the purpose of mechanical, thermal and scanning electronic microscope (SEM) studies, and the water uptake test was also conducted. ASTM D 638-03 standards were followed for the tensile test of the specimen with a 'test speed of 5 mm/min'. The flexural strength test according to ASTM D790 was performed, with a standard test speed of 1.5 mm/mm. The composite specimens were subjected to SEM for the micro-level analysis of failure. An increasing trend in the mechanical strengths, along with an increase in fibre content, was then observed. The SEM image showed the failure of the specimen due to fibre pull-out, as pictured in Figure 12.6.

(Thiruchitrambalam et al. 2009) conducted a study on the hybridising effects of banana and sisal fibres on their respective mechanical properties. Tests like water absorption, tensile, compression, impact and flexural tests were performed on the test specimen. On tensile loading conditions, sisal fibre composites showed brittle failure. Diminished fibre pull-out was observed, and this can possibly be because of

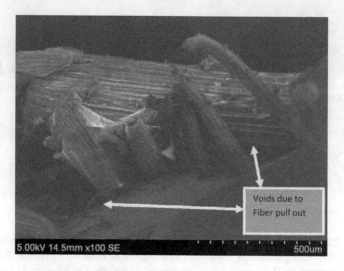

Voids due to
Fiber pull out

5.00kV 14.5mm x100 SE             500um

**FIGURE 12.6** SEM of the Tensile Fracture Surface of 50:50 Banana-Jute Hybrid Composites. (Reprinted from *Composites Part B: Engineering*, 51, Boopalan, M., M. Niranjanaa and M.J. Umapathy Study on the Mechanical Properties and Thermal Properties of Jute and Banana Fibre Reinforced Epoxy Hybrid Composites, 54-57, 2013, with kind permission from Elsevier)

reduced tensile strength, whereas banana fibre showed tensile failure, plastic deformation, lower fibre pull-out and a high elongation percentage. Whereas, the hybrid fibre composites exhibited only partial brittle nature of fracture in sisal fibre composites. The hybrid fibre demonstrated enhanced percentage elongation due to moisture uptake. Moreover, the sisal and hybrid fibres performed well in the compression test. The performances of hybrid and sisal fibres are more or less identical in dry conditions, although sisal fibre lacks performance under high moisture conditions. Under moist conditions, the hybrid composite was able to withstand more stress as compared to the sisal fibre. The hybrid composite also offered better performance in the impact test. Studies show that, due to better adherence between 'fibre and the matrix', the composite of 'volume fraction 0.4' peaked in damping factor value (=tan δ) (Idicula et al. 2005). The same trend was obtained in the static mechanical tests. Figure 12.7(b) depicts the SEM image of the specimen with 0.4 volume; therefore, better adhesion can be surmised, as compared to Figure 7 (b) with 0.2 volume fraction.

(Venkateshwaran et al. 2011) conducted a study on 'banana-sisal reinforced hybrid composites'. It was observed that an increase in the weight ratio of the hybrid enhanced its mechanical strengths. In the water absorption test, it was shown that the specimen with a 50:50 weight proportion took in the least amount of water and thus, had the lowest permeability coefficient. The failure of the composite specimen is due to the pull-out of fibre as shown in Figures 12.8 and 12.9. Figure 12.10 shows poor interfacial bonding in the banana-sisal composite, as compared to the glass fibre composite.

(a)                                       (b)

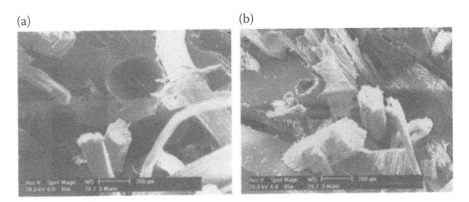

**FIGURE 12.7** SEM Image of 'Banana-Sisal Hybrid Composites at $0.20V_f$ (a × 50) and $0.40V_f$ (b × 50)' having a 'Volume Ratio of Fibres 1:1' Fractured under Tensile Loading. (Reprinted from *Composites Science and Technology*, 65 (7-8), Idicula, Maries, S.K. Malhotra, Kuruvilla Joseph and Sabu Thomas, Dynamic Mechanical Analysis of Randomly Oriented Intimately Mixed Short Banana/Sisal Hybrid Fibre Reinforced Polyester Composite, 1077-1087, 2005, with kind permission from Elsevier)

A detailed analysis of the mechanical characteristics of a hybrid composite of banana, jute and carbon fibre was done by Tamilarasan et al. 2019. Two specimens were prepared, for the purpose of analysis. Specimen A is comprised of a banana-jute hybrid composite, while Specimen B constitutes a composite of banana, jute and carbon fibre. The specimens were also subjected to mechanical tests complying

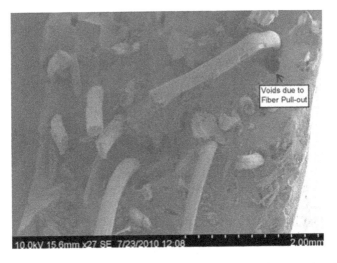

**FIGURE 12.8** SEM Analysis of Banana-Sisal Hybrid Composite, Failed under Tensile Loading. (Reprinted from *Materials and Design*, 32 (7), Venkateshwaran, N., A. ElayaPerumal, A. Alavudeen and M. Thiruchitrambalam, Mechanical and Water Absorption Behaviour of Banana/Sisal Reinforced Hybrid Composites, 4017-4021, 2011, with kind permission from Elsevier)

**FIGURE 12.9**   SEM Analysis of the Banana-Sisal Hybrid Composite, Failed under Flexural Loading. (Reprinted from Materials and Design, 32 (7), Venkateshwaran, N., A. ElayaPerumal, A. Alavudeen and M. Thiruchitrambalam, Mechanical and Water Absorption Behaviour of Banana/Sisal Reinforced Hybrid Composites, 4017-4021, 2011, with kind permission from Elsevier)

with their respective ASTM standards. The bending strength of the banana-jute hybrid composite resulted as 65.35 MPa, and UTS emerged as 21.308 MPa. Meanwhile, the bending strength of the carbon-banana-jute composite was 87.17 MPa, and the ultimate tensile strength is 59.586 MPa. Thus, banana fibre with jute fibre and carbon fibre increases the mechanical characteristics significantly, especially when compared to the original fibre.

Another study indicated that tensile failure can mainly depend on the fibre pull-out (Figure 12.11) (Arthanarieswaran, Kumaravel, and Kathirselvam 2014). As depicted in Figure 12.12, the formation of voids is responsible for poor mechanical characteristics in the banana-sisal composite. However, hybridisation is proven to enhance the strength of 'banana fibre composite' significantly.

The outcomes of hybridising banana-pineapple leaf-glass fibre in different volume fractions were evaluated by Hanafee et al. 2017. The test results proved that the tensile strength of hybrid improved with the subsequent increments in volume fraction, based on studying the best possible volume fraction of banana, and palf resulted as 50%. A further 85% gain in tensile strength was observed upon the addition of a single layer of glass fibre on the banana-palf hybrid.

Alavudeen et al. 2015 fabricated the banana-kenaf hybrid composite. The sample was tested for flexural, impact and tensile strengths, and electron microscopy

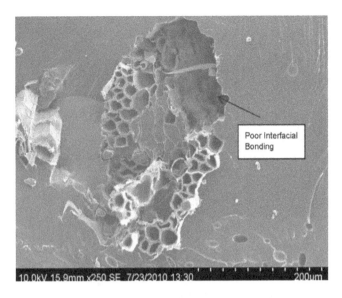

**FIGURE 12.10**   SEM Image of the Banana-Sisal Hybrid Composite, Failed Under Tensile Loading. (Reprinted from Materials and Design, 32 (7), Venkateshwaran, N., A. ElayaPerumal, A. Alavudeen and M. Thiruchitrambalam, Mechanical and Water Absorption Behaviour of Banana/Sisal Reinforced Hybrid Composites, 4017-4021, 2011, with kind permission from Elsevier)

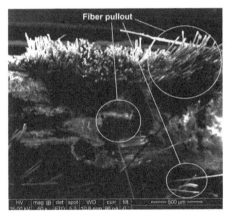

**FIGURE 12.11**   Fibre Pull-Out in Glass-Banana-Sisal Tensile Test Specimen. (Reprinted from Materials and Design, 64, Arthanarieswaran, V.P., A. Kumaravel and M. Kathirselvam, Evaluation of Mechanical Properties of Banana and Sisal Fibre Reinforced Epoxy Composites: Influence of Glass Fibre Hybridisation, 194-202, 2014, with kind permission from Elsevier)

scanning was also performed complying with the respective ASTM standards. In the tensile test, a rise in strength was observed for the banana-kenaf reinforced hybrid, as compared to plain banana and plain kenaf fibres. A similar development was noted in flexural and impact tests. Figure 12.13 and 12.14 show pull-out failure of the hybrid composite.

Senthil Kumar et al. 2016 analysed the vibration properties of the banana-coconut sheath hybrid composite. The composites were fabricated in four varied

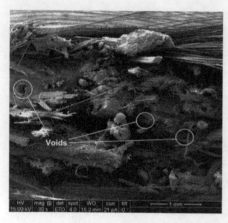

**FIGURE 12.12** Occurrence of Voids in the Glass-Banana-Sisal Tested Specimen. (Reprinted from Materials and Design, 64, Arthanarieswaran, V.P., A. Kumaravel and M. Kathirselvam, Evaluation of Mechanical Properties of Banana and Sisal Fibre Reinforced Epoxy Composites: Influence of Glass Fibre Hybridisation, 194-202, 2014, with kind permission from Elsevier)

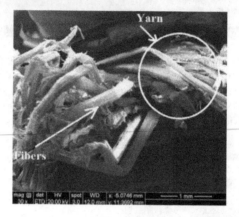

**FIGURE 12.13** Scanning Electron Microscopy Image of 'Randomly Oriented Banana-Kenaf Hybrid Composite' Specimen Showing Failure due to Tensile Loading. (Reprinted from Materials and Design, 66, Alavudeen, A., N. Rajini, S. Karthikeyan, M. Thiruchitrambalam and N. Venkateshwaren, Mechanical Properties of Banana/Kenaf Fibre-Reinforced Hybrid Polyester Composites: Effect of Woven Fabric and Random Orientation, 246-257, 2015, with kind permission from Elsevier)

layering patterns, i.e. carbon-banana-carbon, carbon-carbon-banana, banana-carbon-banana and banana-banana-carbon. The conclusion of the mechanical and vibrational tests showed that the skin eccentric fibre yielded an elevated tensile strength with a greater weight percentage of fibre. Also, the damping value of hybrid composites, that were alkali-treated, was established as higher. Figure 12.15 and Figure 12.16 show the improved fibre-matrix adherence of the CCB-layered composite and the BCC-layered composite.

The physicomechanical characteristics of banana and coir fibres hybrid were investigated by Prasad, Agarwal, and Sinha 2018. The samples were tested for mechanical, thermo-gravimetric, morphological and water intake analysis according to the corresponding ASTM standards. The results of the tests clearly showed that

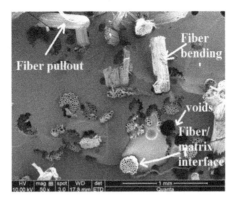

**FIGURE 12.14** Scanning Electron Microscopy Image of 'Randomly Oriented Banana Fibre Composite' Specimen Showing Failure due to Tensile Loading. (Reprinted from Materials and Design, 66, Alavudeen, A., N. Rajini, S. Karthikeyan, M. Thiruchitrambalam and N. Venkateshwaren, Mechanical Properties of Banana/Kenaf Fibre-Reinforced Hybrid Polyester Composites: Effect of Woven Fabric and Random Orientation, 246-257, 2015, with kind permission from Elsevier)

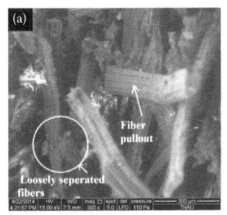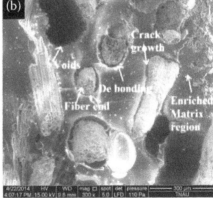

**FIGURE 12.15** SEM of Tensile Tested Coconut Sheath-Banana Hybrid Composites: (a) Untreated CCB and (b) NaOH treated CCB. (Reprinted from Materials and Design, 90, Senthil Kumar, K., I. Siva, N. Rajini, J.T. Winowlin Jappes and S.C. Amico, Layering Pattern Effects on Vibrational Behaviour of Coconut Sheath/Banana Fibre Hybrid Composites, 795-803, 2016, with kind permission from Elsevier)

hybridising coir fibre on banana fibre enhanced its tensile strength, flexural properties, thermal resistance, toughness and water resistance by a compelling margin. From scanning electron microscopy, they could conclude that the addition of coir fibre into the banana fibre composites refined the fibre/matrix interfacial adhesion and reduced the fibre pull-out. On the other hand, Haneefa et al. (2008) analysed the characteristics of the hybrid composite prepared from chopped glass fibre and

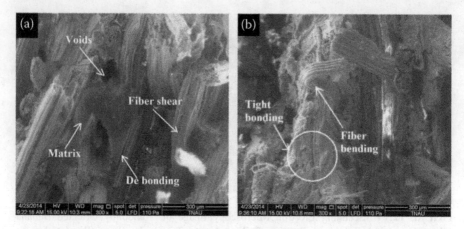

**FIGURE 12.16**   SEM of Tensile Tested Banana-Coconut Sheath Hybrid Composites: (a) Untreated BCC and (b) NaOH. (Reprinted from Materials and Design, 90, Senthil Kumar, K., I. Siva, N. Rajini, J.T. Winowlin Jappes and S.C. Amico, Layering Pattern Effects on Vibrational Behaviour of Coconut Sheath/Banana Fibre Hybrid Composites, 795-803, 2016, with kind permission from Elsevier)

banana fibre. For the fabrication of the composite, raw fibres were chopped, and a solution-mixing process was used to produce the composite. The composite was then prepared using an injection moulding machine, in order to obtain 4 mm cylindrical rods. The specimen was then prepared by compression moulding at 8 MPa and at a temperature of 150°C, with the dimensions 120 × 26.5 × 2.5 mm. The specimens were later subjected to various mechanical tests, and the SEM study was done. The test outcomes showed that the strength, modulus and percentage elongation of the composite depend on the rise in the volume fraction of fibreglass fibre. Zulkafli et al. (2020) studied the impact absorption performance of banana/glass fibre hybrid composites. The composites were arranged in four sets of stacking sequences, i.e. BBB, GGG, BGB, GBG of dimension 250 × 250 mm, using the compression moulding machine. The specimens were subjected to a low viscosity impact test according to ASTM D7136. It was concluded that the addition of glass fibre in the composite improved the impact energy absorption.

Sanjay, Arpitha, and Yogesha (2015) conducted research to be able to characterise banana/e-glass composite on the basis of mechanical properties. To attain the goal of composite fabrication, different laminates of various combinations of glass and banana fibre were prepared using the vacuum bagging method. Firstly, wax was applied to the mat. Then, the six laminates were impregnated with resin isophthalic polyester resin, catalyst and accelerator. This was later followed by sealing the laminates in a vacuum bag. The laminates are left to cure for four hrs in the vacuum bag, at a pressure of 60–70 bar. Additionally, post-curing is done for four hrs at 70 °C. The results revealed that void fractions in the laminates are extremely less because the laminates are healed using the vacuum bagging method. In the tensile test, it was observed that the increasing layers of glass fibres in banana-glass hybrid surge the breaking load, tensile modulus and UTS of the hybrid. A similar trend is apparent in other mechanical tests.

Ramesh et al. (2017) evaluated banana-carbon fibre reinforced hybrid composites for their mechanical properties. Five layers of carbon fibre and banana fibre were positioned in an alternating fashion. Epoxy resin was applied on the surface and distributed with a roller. Finally, the laminate is cured under load for 24 hrs. The size of laminate is 300 × 300 × 5 mm. Tensile, impact, flexural, water intake and morphological tests were conducted on the specimen. The study concluded that the laminate, with a 20:80 fraction of carbon and banana respectively, showed maximum tensile, flexural and impact strength while simultaneously consuming the least amount of water.

Harikrishna et al. (2018) investigated the banana/glass /jute hybrid along with its mechanical characteristics and fabrication. It was observed that the jute-banana-glass hybrid exhibited the most tensile strength amongst the three, and maximum impact resistance, as well as flexural strength, were apparent in a similar trend. The effects of treatment and hybridisation were evaluated by Terwadkar and Walame (2019). Next, mechanical properties were then tested on the specimen. The tests assessed that the hybrid banana and kenaf fibre contain a higher elastic modulus than the chemically treated individual banana and kenaf fibres. Moreover, Pothan et al. (2010) evaluated the dynamic, mechanical and dielectric characteristics of the glass–banana hybrid fibre reinforced-polyester-composites. The tests concluded that the layering orientation has a higher effect on the mechanical and dynamic characteristics of the composite rather than hybridisation.

The hybrid natural fibre composite was tested for dynamic and mechanical characteristics (Imran 2015). The vibrational properties of the specimens were studied, and it was proven that the hybrid composite of banana-sisal fibres provides a maximum damping factor of 3.681 at a volume fraction of b-s 30-10%, which is 2.10 times greater than the banana fibre laminate with a damping factor 1.751; hence, b-s contributes higher vibrational absorbing capacity.

Shireesha et al. (2019) evaluated the mechanical characteristics for bi-directional hybrid natural fibre composite. The composite was prepared by stacking up thin sheets of banana and jute fibres in the orientations 0°, 45°, –45°, 0° respectively and subsequently mixing this with epoxy at varied wt %. The test pieces were subjected to flexural, hardness and tensile tests as per the respective ASTM standards. The results of the hybrid composite at 30 wt % are superior than a pure banana or a pure jute fibre composite. Therefore, the mechanical characteristics will improve due to the mixture of fibres at varied orientations with epoxy at any given weight ratio.

A study on sound insulation properties of the banana-coir-polypropylene hybrid and its effects due to hybridisation was conducted by Singh and Mukhopadhyay (2020). The composite was prepared by the compression moulding process. The outcome of various factors such as the arrangement of fibres, hybridisation, fibre volume fraction, etc. on sound insulation of composites was studied. It was discovered that, as the fibre content increases, the loss of transmission likewise developed in the composite specimen. For finer banana fibres, the variation between transmission losses was higher for maximum and minimum fibre loading for non-hybrid composite. Furthermore, it was also observed that transmission loss is worse in individual fibre composites rather than in hybrid composites with the same fibre loading.

## 12.4 SUMMARY

This chapter reviewed advanced fabrication techniques for banana fibre-based hybrid composites that have the potential to replace traditional composites and possess comparatively better properties. The following conclusions can be drawn based upon the review of past work as discussed in this chapter:

- Banana fibre possesses excellent tensile properties and lightweight characteristics and is, therefore, identified as one of the most promising materials to be used in composites for industrial, automobile and aerospace.
- Hybridising natural fibres with banana fibre has improved the 'strength to weight ratio' of the resulting composite. Increasing the percentage of banana fibres consequentially heightened mechanical strengths while reducing the percentage of water absorption in the hybrid composite, as compared to the pure banana fibre composite. Moreover, hybridising glass fibre with banana fibre improves the strength of the banana fibre composite while simultaneously reducing the moisture uptake.
- Chemical pretreatment of fibres before hybridisation increases the mechanical strength of the banana fibre composite drastically due to the reduced moisture uptake, which, in turn, causes better 'interfacial bonding' between 'the resin and the fibre'.
- Banana fibre hybrid composites are considered suitable for tensile and bending applications, as well as for average load utilities.
- During the fabrication of hybrid composites, banana fibre is mostly paired with sisal, jute and kenaf fibres, while in synthetic fibre, glass fibre is the most common pairing. Even though other natural fibres such as flax, hemp, ramie, palf, among others and synthetic carbon and aramid fibres have proven to enhance the properties of natural fibres upon hybridisation, there is little to no research on the effects of their hybridisation with banana fibre. Nevertheless, this can be a major future research avenue.

## REFERENCES

Alavudeen, A., N. Rajini, S. Karthikeyan, M. Thiruchitrambalam, and N. Venkateshwaren. 2015. "Mechanical Properties of Banana/Kenaf Fiber-Reinforced Hybrid Polyester Composites: Effect of Woven Fabric and Random Orientation." *Materials and Design* 66(PA). Elsevier Ltd: 246–257. doi:10.1016/j.matdes.2014.10.067.

Arthanarieswaran, V. P., A. Kumaravel, and M. Kathirselvam. 2014. "Evaluation of Mechanical Properties of Banana and Sisal Fiber Reinforced Epoxy Composites: Influence of Glass Fiber Hybridization." *Materials and Design* 64. Elsevier Ltd: 194–202. doi:10.1016/j.matdes.2014.07.058.

Asim, Mohammad, Mohammad Jawaid, Naheed Saba, Mohammad Nasir Ramengmawii, and Mohamed Thariq Hameed Sultan. 2017. *Processing of Hybrid Polymer Composites-a*

*Review. Hybrid Polymer Composite Materials: Processing.* Elsevier Ltd. doi:10.1016/
B978-0-08-100789-1.00001-0.

Balakrishnan, P., M. J. John, L. Pothen, M. S. Sreekala, and S. Thomas. 2016. *Natural Fibre
and Polymer Matrix Composites and Their Applications in Aerospace Engineering.*
*Advanced Composite Materials for Aerospace Engineering.* Elsevier Ltd. doi:10.1016/
b978-0-08-100037-3.00012-2.

Balasubramanian, K., Mohamed T. H. Sultan, and N. Rajeswari. 2018. *Manufacturing
Techniques of Composites for Aerospace Applications. Sustainable Composites for
Aerospace Applications.* Elsevier Ltd. doi:10.1016/B978-0-08-102131-6.00004-9.

Bhattacharyya, Debes, Aruna Subasinghe, and Nam Kyeun Kim. 2015. *Natural Fibers: Their
Composites and Flammability Characterizations. Multifunctionality of Polymer
Composites: Challenges and New Solutions.* Elsevier Inc. doi:10.1016/B978-0-323-2
6434-1.00004-0.

Boopalan, M., M. Niranjanaa, and M. J. Umapathy. 2013. "Study on the Mechanical
Properties and Thermal Properties of Jute and Banana Fiber Reinforced Epoxy Hybrid
Composites." *Composites Part B: Engineering* 51. Elsevier Ltd: 54–57. doi:10.1016/
j.compositesb.2013.02.033.

Dai, D., and M. Fan. 2013. *Wood Fibres as Reinforcements in Natural Fibre Composites:
Structure, Properties, Processing and Applications. Natural Fibre Composites:
Materials, Processes and Applications.* Woodhead Publishing Limited. doi:10.1533/
9780857099228.1.3.

Ebnesajjad, Sina, and Sina Ebnesajjad. 2003. "PART II 7 Injection Molding." *Melt
Processible Fluoroplastics* 151–193. doi:10.1016/B978-1-884207-96-9.50010-2.

El-Hajjar, R., H. Tan, and K. M. Pillai. 2013. "Advanced Processing Techniques for Composite
Materials for Structural Applications." *Developments in Fiber-Reinforced Polymer
(FRP) Composites for Civil Engineering,* 54–76. doi:10.1533/9780857098955.1.54.

Elkington, M., D. Bloom, C. Ward, A. Chatzimichali, and K. Potter. 2015. "Hand Layup:
Understanding the Manual Process." *Advanced Manufacturing: Polymer and
Composites Science* 1 (3): 138–151. doi:10.1080/20550340.2015.1114801.

Erden, Seçkin, and Kingsley Ho. 2017. "Fiber Reinforced Composites." *Fiber Technology
for Fiber-Reinforced Composites* 51–79. doi:10.1016/B978-0-08-101871-2.00003-5.

Faruk, Omar, Andrzej K. Bledzki, Hans Peter Fink, and Mohini Sain. 2012. "Biocomposites
Reinforced with Natural Fibers: 2000-2010." *Progress in Polymer Science* 37 (11).
Elsevier Ltd: 1552–1596. doi:10.1016/j.progpolymsci.2012.04.003.

Fu, Shao Yun, Bernd Lauke, Edith Mäder, Chee Yoon Yue, Xiao Hu, and Yiu Wing Mai.
2001. "Hybrid Effects on Tensile Properties of Hybrid Short-Glass-Fiber-and Short-
Carbon-Fiber-Reinforced Polypropylene Composites." *Journal of Materials Science* 36
(5): 1243–1251. doi:10.1023/A:1004802530253.

Hanafee, Z. M., A. Khalina, M. Norkhairunnisa, Z. Edi Syams, and K. E. Liew. 2017. "The
Effect of Different Fibre Volume Fraction on Mechanical Properties of Banana/
Pineapple Leaf (PaLF)/Glass Hybrid Composite." *AIP Conference Proceedings* 1885.
doi:10.1063/1.5002339.

Haneefa, Anshida, Panampilly Bindu, Indose Aravind, and Sabu Thomas. 2008. "Studies on
Tensile and Flexural Properties of Short Banana/Glass Hybrid Fiber Reinforced
Polystyrene Composites." *Journal of Composite Materials* 42 (15): 1471–1489. doi:1
0.1177/0021998308092194.

Harikrishna, M., K. Ajeeth, S. Ranganatha, and C. Thiagarajan. 2018. "Fabrication and
Mechanical Properties of Hybrid Natural Fiber Composites (Jute/Banana/Glass)."
*International Journal of Pure and Applied Mathematics* 119 (15): 685–696.

Huda, M. S., L. T. Drzal, D. Ray, A. K. Mohanty, and M. Mishra. 2008. "Natural-Fiber
Composites in the Automotive Sector." *Properties and Performance of Natural-Fibre
Composites* 221–268. doi:10.1533/9781845694593.2.221.

Idicula, Maries, S. K. Malhotra, Kuruvilla Joseph, and Sabu Thomas. 2005. "Dynamic Mechanical Analysis of Randomly Oriented Intimately Mixed Short Banana/Sisal Hybrid Fibre Reinforced Polyester Composites." *Composites Science and Technology* 65 (7–8): 1077–1087. doi:10.1016/j.compscitech.2004.10.023.

Imran, Khalid. 2015. "Experimental Investigation of Mechanical and Dynamic Characteristics of Hybrid Natural Fiber." *International Journal of Innovations in Engineering Research and Technology [IJIERT]* 2 (8): 1–7.

Ismail, N. F., A. B. Sulong, N. Muhamad, D. Tholibon, M. K.F. MdRadzi, and W. A.S. WanIbrahim. 2015. "Review of the Compression Moulding of Natural Fiber-Reinforced Thermoset Composites: Material Processing and Characterisations." *Pertanika Journal of Tropical Agricultural Science* 38 (4): 533–547.

Jamir, Mohammad R. M., Mohammad S. A. Majid, and Azduwin Khasri. 2018. *Natural Lightweight Hybrid Composites for Aircraft Structural Applications. Sustainable Composites for Aerospace Applications.* Elsevier Ltd. doi:b10.1016/B978-0-08-1 02131-6.00008-6.

Jones, D., G. O. Ormondroyd, S. F. Curling, C. M. Popescu, and M. C. Popescu. 2017. "Chemical Compositions of Natural Fibres." In *Advanced High Strength Natural Fibre Composites in Construction,* Edited by Fan and Fu. Woodhead Publishing. doi:10.101 6/B978-0-08-100411-1.00002-9.

Joseph, Seena, M. S. Sreekala, Z. Oommen, P. Koshy, and Sabu Thomas. 2002. "A Comparison of the Mechanical Properties of Phenol Formaldehyde Composites Reinforced with Banana Fibres and Glass Fibres." *Composites Science and Technology* 62 (14): 1857–1868. doi:10.1016/S0266-3538(02)00098-2.

Jústiz-Smith, Nilza G., G. Junior Virgo, and Vernon E. Buchanan. 2008. "Potential of Jamaican Banana, Coconut Coir and Bagasse Fibres as Composite Materials." *Materials Characterization* 59 (9): 1273–1278. doi:10.1016/j.matchar.2007.10.011.

Kılınç, Ahmet Çağri, Cenk Durmuşkahya, and M. Özgür Seydibeyoğlu. 2017. "Natural Fibers." In *Fiber Technology for Fiber-Reinforced Composites,* Edited by Seydibeyoglu et al. 209–235. doi:10.1016/B978-0-08-101871-2.00010-2.

Kim, Y. K. 2012. *Natural Fibre Composites (NFCs) for Construction and Automotive Industries. Handbook of Natural Fibres.* Woodhead Publishing Limited. doi:10.1533/ 9780857095510.2.254.

Kretsis, G. 1987. "A Review of the Tensile, Compressive, Flexural and Shear Properties of Hybrid Fibre-Reinforced Plastics." *Composites* 18 (1): 13–23. doi:10.1016/0010-4361 (87)90003-6.

Kulkarni, A. G., K. G. Satyanarayana, P. K. Rohatgi, and Kalyani Vijayan. 1983. "Mechanical Properties of Banana Fibres (Musa Sepientum)." *Journal of Materials Science* 18 (8): 2290–2296. doi:10.1007/BF00541832.

Kureemun, Umeyr, M. Ravandi, L. Q. N. Tran, W. S. Teo, T. E. Tay, and H. P. Lee. 2018. "Effects of Hybridization and Hybrid Fibre Dispersion on the Mechanical Properties of Woven Flax-Carbon Epoxy at Low Carbon Fibre Volume Fractions." *Composites Part B: Engineering* 134. Elsevier Ltd: 28–38. doi:10.1016/j.compositesb.2017.09.035.

Liu, S.-J. 2012. *Injection Molding in Polymer Matrix Composites. Manufacturing Techniques for Polymer Matrix Composites (PMCs).* Woodhead Publishing Limited. doi:10.1533/ 9780857096258.1.13.

Madhukiran, J, S. Srinivasa Rao, and S. Madhusudan. 2013. "Tensile And Hardness Properties Of Banana/Pineapple Natural Fibre Reinforced Hybrid Composites." *International Journal of Engineering Research & Technology (IJERT)* 2 (7): 1260–1264.

Mallick, P. K. 2010. *Thermoset-Matrix Composites for Lightweight Automotive Structures. Materials, Design and Manufacturing for Lightweight Vehicles.* Woodhead Publishing Limited. doi:10.1533/9781845697822.1.208.

Navaneethakrishnan, G., V. Selvam, and S. Julyes Jaisingh. 2015. "Development and Mechanical Studies of Glass/Banana Fiber Hybrid Reinforced Silica Nano Particles with Epoxy Bio-Nanocomposites." *Journal of Chemical and Pharmaceutical Sciences* 7 (7): 197–199.

Oksman, Kristiina. 2001. "High Quality Flax Fibre Composites Manufactured by the Resin Transfer Moulding Process." *Journal of Reinforced Plastics and Composites* 20 (7): 621–627. doi:10.1177/073168401772678634.

Panthapulakkal, S., L. Raghunanan, M. Sain, B. Kc, and J. Tjong. 2017. "Natural Fiber and Hybrid Fiber Thermoplastic Composites: Advancements in Lightweighting Applications." In *Green Composites: Waste and Nature-Based Materials for a Sustainable Future*, Second Edition, Edited by Baillie and Jayasinghe. Woodhead Publishing. doi:10.1016/B978-0-08-100783-9.00003-4.

Pothan, Laly A., Chandy N. George, Maya Jacob John, and Sabu Thomas. 2010. "Dynamic Mechanical and Dielectric Behavior of Banana-Glass Hybrid Fiber Reinforced Polyester Composites." *Journal of Reinforced Plastics and Composites* 29 (8): 1131–1145. doi:10.1177/0731684409103075.

Prasad, Nirupama, Vijay Kumar Agarwal, and Shishir Sinha. 2018. "Hybridization Effect of Coir Fiber on Physico-Mechanical Properties of Polyethylene-Banana/Coir Fiber Hybrid Composites." *Science and Engineering of Composite Materials* 25 (1): 133–141. doi:10.1515/secm-2015-0446.

Raji, Marya, Hind Abdellaoui, Hamid Essabir, Charles Amani Kakou, Rachid Bouhfid, and Abou El Kacem Qaiss. 2018. "Prediction of the Cyclic Durability of Woven-Hybrid Composites." In Mohammad Jawaid, Mohamed Thariq and Naheed Saba (Eds) *Durability and Life Prediction in Biocomposites, Fibre-Reinforced Composites and Hybrid Composites*. Woodhead Publishing. pp. 27–62. doi:10.1016/B978-0-08-102290-0.00003-9.

Ramesh, Manickam. 2018. *Hemp, Jute, Banana, Kenaf, Ramie, Sisal Fibers. Handbook of Properties of Textile and Technical Fibres*. Elsevier Ltd. doi:10.1016/B978-0-08-1012 72-7.00009-2.

Ramesh, Manickam, Ravi Logesh, Manivannan Manikandan, Nithyanandam Sathesh Kumar, and Damodaran Vishnu Pratap. 2017. "Mechanical and Water Intake Properties of Banana-Carbon Hybrid Fiber Reinforced Polymer Composites." *Materials Research* 20 (2): 365–376. doi:10.1590/1980-5373-MR-2016-0760.

Samivel, P., and Ramesh Babu. 2013. "Mechanical Behavior of Stacking Sequence in Kenaf and Banana Fiber Reinforced-Polyester Laminate." *International Journal of Mechanical Engineering and Robotics Research* 2 (4): 348–360.

Sanjay, M. R., G. R. Arpitha, and B. Yogesha. 2015. "Study on Mechanical Properties of Natural - Glass Fibre Reinforced Polymer Hybrid Composites: A Review." *Materials Today: Proceedings* 2 (4–5). Elsevier Ltd.: 2959–2967. doi:10.1016/j.matpr.2015.07.2 64.

Satyanarayana, K. G., K. Sukumaran, P. S. Mukherjee, C. Pavithran, and S. G. K. Pillai. 1990. "Natural Fibre-Polymer Composites." *Cement and Concrete Composites* 12 (2): 117–136. doi:10.1016/0958-9465(90)90049-4.

Senthil Kumar, K., I. Siva, N. Rajini, J. T. Winowlin Jappes, and S. C. Amico. 2016. "Layering Pattern Effects on Vibrational Behavior of Coconut Sheath/Banana Fiber Hybrid Composites." *Materials and Design* 90. Elsevier B.V.: 795–803. doi:10.1016/ j.matdes.2015.11.051.

Shireesha, Yegireddi, Bade Venkata Suresh, M. V. A. Raju Bahubalendruni, and Govind Nandipati. 2019. "Experimental Investigation on Mechanical Properties of Bi-Directional Hybrid Natural Fibre Composite (HNFC)." *Materials Today: Proceedings* 18. Elsevier Ltd.: 165–174. doi:10.1016/j.matpr.2019.06.290.

Singh, Vikas Kumar, and Samrat Mukhopadhyay. 2020. "Studies on the Effect of Hybridization on Sound Insulation of Coir-Banana-Polypropylene Hybrid Biocomposites." *Journal of Natural Fibers* 00 (00). Taylor & Francis: 1–10. doi:10.1080/15440478.2020.1745116.

Tamilarasan, U., P. Kishore Kumar, A. Mohamed Shafeeq, S. Vairava Sundaram, and U. Pandiyan. 2019. "Study of Mechanical Properties of Banana-Jute Fibre Epoxy Composite with Carbon Fibre." *Composites Part B Engineering* 6 (6): 160–167.

Tanzi, Maria Cristina, Silvia Farè, and Gabriele Candiani. 2019. "Manufacturing Technologies." *Foundations of Biomaterials Engineering* 149. Elsevier Ltd.: 137–196. doi:10.1016/B978-0-08-101034-1.00003-7.

Tatara, Robert A. 2011. *Compression Molding. Applied Plastics Engineering Handbook.* Elsevier. doi:10.1016/B978-1-4377-3514-7.10017-0.

Terwadkar, Aniket A., and M. V. Walame. 2019. "Mechanical Properties of Banana Fabric and Kenaf Fiber Reinforced Epoxy Composites: Effect of Treatment and Hybridization." *International Journal of Innovative Technology and Exploring Engineering* 8 (10): 1870–1874. doi:10.35940/ijitee.J9230.0881019.

Thiruchitrambalam, M., A. Alavudeen, A. Athijayamani, N. Venkateshwaran, and A. Elaya Perumal. 2009. "Improving Mechanical Properties of Banana/Kenaf Polyester Hybrid Composites Using Sodium Laulryl Sulfate Treatment." *Materials Physics and Mechanics* 8 (2): 165–173.

Van Hattum, F. W.J., F. Regel, and M. Labordus. 2011. "Cost Reduction in Manufacturing of Aerospace Composites." *Plastics, Rubber and Composites* 40 (2): 93–99. doi:10.11 79/174328911X12988622801052.

Varghese, Anish M., and Vikas Mittal. 2017. "Surface Modification of Natural Fibers." In *Biodegradable and Biocompatible Polymer Composites: Processing, Properties and Applications.* Edited by N. G. Shimpi. Woodhead Publishing. doi:10.1016/B978-0-08-100970-3.00005-5.

Venkateshwaran, N., A. ElayaPerumal, A. Alavudeen, and M. Thiruchitrambalam. 2011. "Mechanical and Water Absorption Behaviour of Banana/Sisal Reinforced Hybrid Composites." *Materials and Design* 32 (7). Elsevier Ltd: 4017–4021. doi:10.1016/j.matdes.2011.03.002.

Zulkafli, Norizzati, Sivakumar Dhar Malingam, Siti Hajar Sheikh Md Fadzullah, and Nadlene Razali. 2020. "Quasi and Dynamic Impact Performance of Hybrid Cross-Ply Banana/Glass Fibre Reinforced Polypropylene Composites." *Materials Research Express* 6 (12): 125344. doi:10.1088/2053-1591/ab5f8c.

# 13 Effect of Microstructure and Post-Processing on the Corrosion Behaviour of Metal Additive Manufactured Components

*P. K. Diljith, A. N. Jinoop, C. P. Paul, and K. S. Bindra*

## CONTENTS

## 13.1 INTRODUCTION

Additive manufacturing (AM) is the process of manufacturing components layer-by-layer directly from its 3D model as opposed to subtractive and formative manufacturing technologies. AM techniques possess the ability to process a wide range of materials in the desired composition, density and functionalities in order to build sound components with intricate geometry from a predesigned model. Metal additive manufacturing (MAM) extended the applications of AM from prototyping to industrial manufacturing. MAM can be classified as wire arc additive manufacturing (WAAM), electron beam melting (EBM) and laser additive manufacturing (LAM) based on the heat source used and as powder bed fusion (PBF) and directed energy deposition (DED) based on the material feeding method.

WAAM is an extension of conventional welding used to build 3D components. As shown in Figure 13.1(a), the WAAM setup is equipped with an electric heat source, a motion guidance system and a wire-based direct feedstock input. The electric arc melts the feedstock and is fed in wire form. WAAM is widely accepted due to its high deposition rate, low equipment cost, low wastage and resulting eco-friendliness. Manufacturing time and cost are substantially reduced by using WAAM. However, due to residual stress, distortion and the presence of porosity, the tensile strength and fatigue life of WAAM-built samples are affected. The poor geometrical accuracy of WAAM-built components necessitates its machining prior to fabrication. WAAM is employed in the large-scale production of aerospace, automotive and rapid tooling industries (Wu, Pan, Ding et al. 2018; Johnnieew Li et al. 2019).

The laser powder bed fusion process (LPBF), commercially known as selective laser melting (SLM) was developed to obtain fully dense parts with comparable mechanical properties to their conventional counterparts. As shown in Figure 13.1(b), metal powder spread on a build plate is melted by a laser beam, as guided by a 3D model, in order to fuse with the substrate or previous layer. LPBF can build complex parts and thin sections with detailed internal features, undercuts and intricate profiles. LPBF parts show superior surface finish and have comparable material properties to their conventional counterparts. However, the process is inapplicable in printing parts of large volumes. The ductility of the manufactured product is lower in LPBF due to a higher dislocation density. The employment of online monitoring and closed-loop control is now widely researched to increase the repeatability and reproducibility of the process (Yakout, Elbestawi, and Veldhuis 2018; Nayak et al. 2020).

As opposed to LPBF, the laser directed energy deposition (LDED) system uses a high-intensity laser beam in generating a melt pool in the substrate or previous build layer and deposits the metal into the melt pool. As shown in Figure 13.1(c), a nozzle,

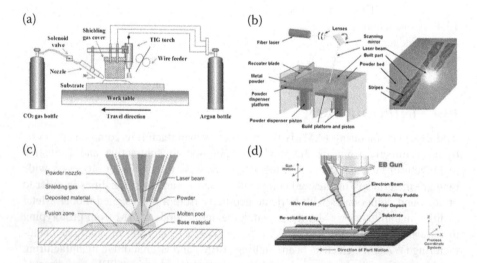

**FIGURE 13.1** Schematic Diagram of (a) WAAM (Wu, Pan, Li et al. 2018), (b) LPBF (Criales et al. 2017), (c) LDED (Graf et al. 2013) and (d) EBM (Fedorov et al. 2019) Based MAM processes.

coaxial to the laser source, delivers the powder in a stream of argon gas, which is also used as the shielding gas. Even though the surface finish and geometrical intricacy are low, near net-shaped multi-material components and functionally graded components can be built using DED at a higher build rate and with less powder. The LDED-built samples have high density and tensile strength as compared to the conventional samples. LDED is used in the repair and remanufacturing of tooling parts and the aerospace and petrochemical industries (Jinoop, Paul, and Bindra 2019; Yakout, Elbestawi, and Veldhuis 2018; Jinoop et al. 2019; Yadav et al. 2019).

Electron beam melting (EBM) melts the metal using an electron beam. Initially, a high-intensity electron beam, at a higher scan speed, preheats the powder. Next, a low-intensity electron beam, at a lower scan speed, melts the powder, as shown in Figure 13.1(d). Preheating not only reduces the beam energy needed for the deposition but also reduces residual stresses developed in the melted samples. EBM is performed in a vacuum in order to conserve the energy of the electron beam. Support structures are not required for structural support, but instead are present for heat transfer. EBM process can achieve build rates as high as 80 cm$^3$/hr. The vacuum reduces the oxidation, air entrapment and the presence of impurities but makes the process complex as well as expensive. The processing powder must be electrically conductive. The dimensional accuracy and the surface finish of the EBM built samples are less than that of LPBF based parts. The EBM counterpart of LDED is called electron beam-directed energy deposition (EB-DED). Unlike LDED, wire feed is more advisable over powder feed in EB-DED due to the powder flow restrictions in a vacuum. EB-DED is desirable in processing metals with higher reflectivity, because the amount of energy absorbed is not a function of the absorptivity of the material (Yakout, Elbestawi, and Veldhuis 2018; Fedorov et al. 2019).

The properties of the MAM components depend on the process parameter selection for deposition. Every point in the additive manufactured sample has a complex thermal history of rapid melting and solidification, as well as several cycles of heating and cooling. These variations in cooling rates can affect the microstructure and, in turn, the thermal, material, mechanical, fatigue and electrochemical properties of the samples. The major parameters and their effects are presented in Table 13.1 (Yakout, Elbestawi, and Veldhuis 2018).

## TABLE 13.1
## Process Parameters in MAM

| Parameter | Effect of Increasing the Parameter |
| --- | --- |
| beam power | greater solidification time, coarser grain, evaporation of feedstock at extreme laser powers, porosity in the samples, larger heat affected zone and distortion, more residual stress, increasing of density up to a point and then decreases. |
| scan speed | lower melting of substrate, lack of fusion porosity, finer grains |
| material feed rate | increase in catchment efficiency, increase in deposition rate, spatter formation |

## 13.2 CORROSION BEHAVIOUR

The formation of undesirable compounds on the surface of materials due to chemical reactions with foreign elements present in the atmosphere is called corrosion. In the past, corrosion was measured by an immersion test, but this was time-consuming and inaccurate. Electrochemical corrosion testing is an advanced testing method that can provide insight into the corrosion rate, passivation behaviour and pitting behaviour of the sample under investigation (Thirumalaikumarasamy, Shanmugam, and Balasubramanian 2014).

The electrochemical testing of corrosion is based on the analogy of anodic dissolution in an electrochemical cell to metal removal in corrosion, and the procedure is as follows. The sample is exposed to the environment and connected to a reference electrode. A potentiostat supplies the potential and records the current output to give various plots. The equilibrium potential, open circuit potential (OCP) or corrosion potential (Ecorr), which is the affinity of the metal to undergoing corrosion, can be measured using an open-circuit potential curve. The Tafel plot, as shown in Figure 13.2(a), presents the corrosion current ($I_{corr}$) as corresponding to the Ecorr. The corrosion rate (CR) can be calculated from $I_{corr}$ using Equation (13.1), which is derived from Faraday's law of electrolysis:

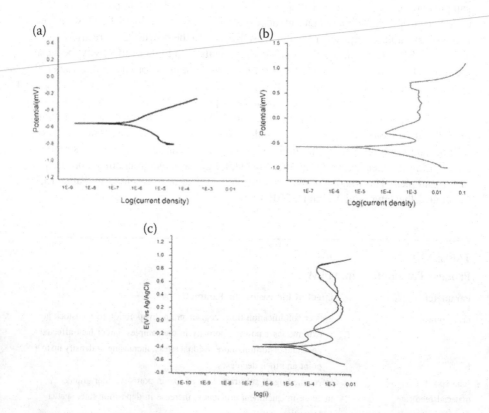

FIGURE 13.2    (a) Tafel Plot (b) Potentiodynamic Plot and (c) Cyclic Potentiodynamic plot.

$$C. R. (mpy) = \frac{0.13 I_{CORR}(E. W.)}{d} \qquad (13.1)$$

Where,

EW = equivalent weight of the sample under study (grams)

$d$ = density of the sample under study (g/cm$^{2)}$

$I_{CORR}$ = corrosion current density, $\mu A/cm^2$

Metals are passivated by forming a protective oxide layer on the corroding surface by likewise reacting with the corroding solution. The potentiodynamic curve, as shown in Figure 13.2(b), reveals the passivation behaviour of the samples. The formation of pits on the corroding surface, which is fatal for the parts, can be predicted using a cyclic potentiodynamic curve, as illustrated in Figure 13.2(c).

MAM techniques are becoming popular in the manufacturing sector. Many additive manufactured parts are widely employed in extremely corrosive environments, such as in the aerospace, aviation, petrochemical and nuclear industries. Therefore, it is necessary to deeply understand the electrochemical behaviour of MAM samples. This paper reviews the most commonly used metallic materials manufactured by MAM along with their electrochemical behaviour, such as nickel superalloys and titanium-based alloys (Buchanan and Stansbury 2012; Rao 2004).

## 13.3 CORROSION BEHAVIOUR OF MAM-BUILT COMPONENTS

### 13.3.1 NICKEL SUPERALLOYS

Nickel superalloys are a favourite group of materials in the manufacturing industry. Because of the difficulty in its machining and the corresponding employability of MAM for the production of intricate profiles applied in corrosive environments, such as turbine blades, the investigations on the corrosion behaviour of nickel superalloys built by different MAM processes are vital. Various studies are conducted in order to understand the corrosion behaviour in IN718, one of the most widely used nickel superalloy that is produced by MAM. LPBF built IN718 samples are studied and compared to conventional and post-processed samples by various authors across the globe (Luo et al. 2019; Popovich et al. 2017).

The electrochemical behaviour of IN718 built by LPBF and its casted counterpart is investigated by Raj et al. 2019. The IN718 sample was built using a laser with a beam diameter of 0.08 mm, at a laser power of 285 W and a scan speed of 970 mm/s. The LPBF, as well as cast samples, were solutionized at 1,100°C for two hours and age-hardened at 845°C for 24 hours. The as-built and heat-treated samples are studied for corrosion behaviour in 1M $H_2SO_4$, with reference to a standard calomel electrode at a scan speed of 1 mV/sec. The order of corrosion resistance obtained after experimentation is: Heat-treated LPBF > as-built LPBF > as-built cast > heat-treated cast alloy. It was observed that the segregation of alloying elements, like Molybdenum, Titanium and Niobium, are higher in LPBF samples after heat treatment. The reaction of these elements heightened the corrosion resistance of heat-treated LPBF samples. The corrosion rate of commercial samples increased with precipitation of 54% of iron.

One of the methods used to enhance the corrosion resistance of MAM-built IN718 is by doping other elements or reinforcements, like TiC (Kong et al. 2019), Re (Kurzynowski et al. 2017), etc. The effect of Rhenium (Re) addition on IN718 samples by LAM and change in corrosion behaviour with build orientation is studied by Majchrowicz et al. (2018). The specimens are built in two build orientations (0° and 90°) using LPBF by mixing 0, 2, 4 and 6 wt % of Re powder, using commercially available IN718 powders. The electrochemical studies were performed at a scan rate of 0.2 mV/sec in 0.1 M $Na_2SO_4$ and 0.1 M NaCl solutions at room temperature using a calomel reference electrode. The as-built IN718 samples with no Re addition show an Icorr value of 0.200 $\mu A/cm^2$, 0.150 $\mu A/cm^2$ in 0.1 M $Na_2SO_4$ and 0.1 M NaCl respectively. The Icorr values of their 90°-oriented counterparts in the respective solutions are one order less. The reduction in corrosion current, as well as the positive shift in corrosion potential, show that the corrosion resistance of IN718 elevated with the addition of Re alloys in both of the build orientations. It was also observed that the XY plane at 90° orientation displayed more corrosion resistance than XZ plane at 0° orientation. The presence of a larger fraction of overlapping area in the XZ plane could be the reason for the increased corrosion rate.

The anisotropy in MAM samples leads to different corrosion rates in varying sections. The effect of the microstructure and heat treatment on corrosion behaviour of IN718 built by DED techniques are, thus, reported (Zhang et al. 2020).

The vertical and horizontal sections of the samples deposited using LDED IN718 was investigated for electrochemical behaviour (Guo, Lin, Li et al. 2018). Gas atomised IN718 powders of 45 - 150 μm size are deposited using laser power – 4 kW, scan speed – 15 mm/s, powder feed rate – 30 g/min, spot diameter – 5 mm, overlap – 50% and layer thickness – 0.9 mm. The potentiodynamic tests were performed at a potential scan rate of 1 mV/sec using a saturated calomel electrode in a 10 wt % $NaNO_3$ at a temperature of 25°C. The microstructure analysis of the LDED deposited IN718 samples, as shown in Figure 13.3, reveals equiaxed grains in the horizontal section. Coarse columnar epitaxial grains, which extend vertically over multiple layers are observed in the vertical section. The passivation current obtained for both the horizontal and vertical sections are identical. This leads to the fact that the microstructural distinctions along the different sections do not affect corrosion rate. However, the potentiodynamic plot as shown in Figure 13.3(e) shows that the corrosion potential of the vertical section (−0.78 V) is less compared to the horizontal section (−0.55 V). This can be due to the fact that the rapidly dissolving γ matrix phase is formed more in the vertical plane rather than on the horizontal plane.

The effect of anisotropy on corrosion behaviour and the influence of different post-processing techniques is studied on wire arc additive manufactured samples (Shen et al. 2018, 2019). Heat-treated wire arc additive manufactured (WAAM) IN718 and its wrought counterpart are compared using electrochemical methods by Zhang and Ojo 2020. The depositions were made by TIG-based WAAM using a 0.787 mm diameter filler wire with an applied current of 100 A, arc length of 3.0 mm, wire feed rate of 0.5 m/min and scanning speed of 0.13 m/min. Vacuum induction melting followed by homogenising, forging and rolling at elevated

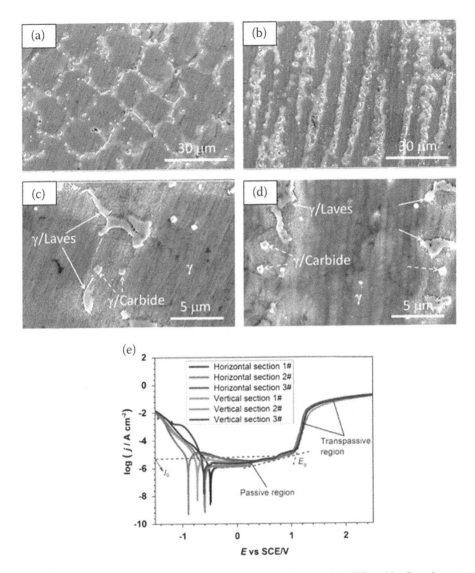

**FIGURE 13.3**    SEM Micrographs of the Dendrites on (a) HS and (b) VS and its Constituent Phases on (c) HS and (d) VS etched with $CuSO_4$-$H_2SO_4$-HCl solution (e) potentiodynamic polarisation curves for the HS and VS of relieved-stress annealing samples in $NaNO_3$ solution (Reprinted from Corrosion Science, 132, Guo, Pengfei, Xin Lin, Jiaqiang Li, Yufeng Zhang, Menghua Song and Weidong Huang, Electrochemical Behaviour of Inconel 718 Fabricated by Laser Solid Forming on Different Sections, 79–89, 2018, with kind permission from Elsevier).

temperature is used to produce the wrought alloy. The WAAM IN718 and the wrought alloy are subjected to identical solution treatment and ageing. Potentiodynamic polarisation tests at 1 mV/s scan rate are conducted in 1M $HNO_3$ and 1M $H_2SO_4$ with saturated calomel electrode as the reference electrode. The

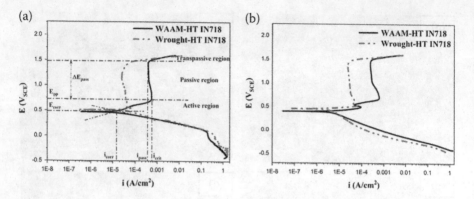

**FIGURE 13.4** Comparison of Potentiodynamic Curves of WAAM-HT IN718 Specimen and Wrought-HT specimen in (a) 1M $HNO_3$ solution and (b) 1M $H_2SO_4$ solution (Reprinted from *Journal of Alloys and Compounds*, 829, Zhang, L.N., and O.A. Ojo, Corrosion Behaviour of Wire Arc Additive Manufactured Inconel 718 Superalloy, 154455, 2020, with kind permission from Elsevier).

tests reveal that the corrosion resistance of the WAAM-built sample is far less than that of the heat-treated conventional built sample in 1M $HNO_3$. The potential range in which the sample stays passivated is smaller in heat-treated WAAM IN718 rather than in wrought IN718, which indicates that that the passive layer formed on it is less stable compared to its counterpart. Likewise, a similar trend is shown in 1M $H_2SO_4$, as presented in Figure 13.4. The corrosion resistance can be related to the stability of the passive layer, and it was observed that the heat-treated WAAM IN718 contained more nickel oxide in its passive layer. Nickel oxide is more porous and less protective than chromium oxide. The formation of nickel oxide in a larger amount can be attributed to the difference in microstructure of the heat-treated WAAM IN718 specimen and the heat-treated wrought samples. Another reason is the increased activity of Ni over Cr due to the reduction of Nb concentration in the matrix since the Laves and delta phase formation during the solidification of WAAM built sample takes up the Nb in the matrix.

## 13.3.2 TITANIUM ALLOYS

Titanium is one of the most popular alloys in the manufacturing industry due to its superior strength to weight ratio and corrosion resistance. The effect of the MAM process parameters on the corrosion behaviour of titanium parts is well explored in the literature (Jiaqiang Li et al. 2019; Chiu et al. 2018).

One of the recent studies (Mahamood and Akinlabi 2018) investigated the influence of deposition parameters on the corrosion behaviour of LDED-built Ti6Al4V. Gas atomised Ti6Al4V powders of the size 150–250 μm are deposited at a scan speed of 0.005 m/s using an Nd-YAG laser with a spot size of 2 mm in the argon gas environment. The powder is fed at a rate of 1.44 g/min in a stream of argon gas supplied at a rate of 4 l/min. The laser power is varied from 0.8–3.0 kW, and the obtained samples are tested for their corrosion resistance in reference to the AgCl electrode in 3.5% NaCl aqueous solution. The results indicate that an increase in laser power will lead to a decrease in the corrosion rate. The microstructure of the conventionally manufactured substrate Ti6Al4V is said to contain alpha and beta grains, whereas the LDED-built surface has epitaxial grains with columnar growth. With the rise in laser power, coarser grains with reduced microhardness are produced. The Widmanstätten alpha is established as increasing with laser power. Due to these reasons, the corrosion resistance is observed to increase with laser power as well.

Similar to the studies conducted in LDED-built samples, the effect of anisotropy of LPBF-built titanium alloys and their heat-treated counterparts, on their corrosion behaviour is studied in various solutions (Chen et al. 2017; Chandramohan et al. 2017; Xu et al. 2017).

Dai et al. (2016) studied the change in corrosion behaviour of LPBF-built Ti6Al4V in different planes. The samples were built using 200 W Yb: YAG fibre laser with an 80 μm spot diameter. Deposition was accomplished in a zigzag pattern with a 90° direction change after every layer and at a scan speed of 1,250 mm/s. To rule out the effect of surface defects, the bulk samples are sectioned along a build plane or the XY plane and built height or the XZ plane, in which cross-sections are illustrated in Figure 13.5, and the cut surface is tested for corrosion in 3.5 wt% NaCl solution and 1M HCl, with respect to standard calomel electrode, at a scan rate of 0.1667 mV/s. The sample also underwent an immersion test by dipping in 1M HCl. The electrochemical tests show that the corrosion resistance of XY-plane is superior to that in the XZ-plane in both of the tested media, which can be attributed to the fraction of α′-phase and β-Ti phase in the microstructure of these planes. Figure 13.6 shows that β-Ti phase forms a protective oxide layer, which is stable than that formed by metastable α′ martensite. The XY-plane has more β-Ti phase and less α′ martensite in its microstructure than XZ-plane. The varied

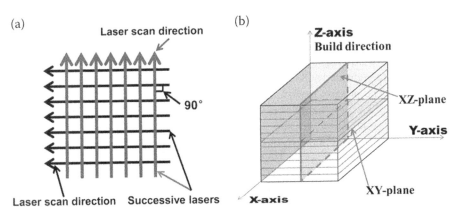

**FIGURE 13.5** Schematic Representation of (a) Laser Scan Direction and the Angle between the Successive Layers in the Process of LPBF and (b) 3D Diagram of LPBF-Built Ti-6Al-4V Alloy and the Planes under Consideration (Reprinted from *Corrosion Science*, 111, Dai, Nianwei, Lai-Chang Zhang, Junxi Zhang, Xin Zhang, Qingzhao Ni, Yang Chen, Maoliang Wu and Chao Yang, Distinction in Corrosion Resistance of Selective Laser Melted Ti-6Al-4V Alloy on Different Planes, 703–710, 2016, with kind permission from Elsevier).

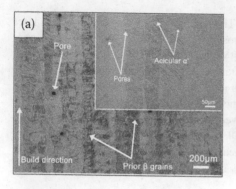
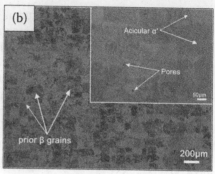

**FIGURE 13.6** Microstructure of (a) XZ-Plane and (b) XY-Plane of LPBF-Built Ti-6Al-4V Sample. Magnified images corresponding to each plane are shown in the insets. (Reprinted from *Corrosion Science*, 111, Dai, Nianwei, Lai-Chang Zhang, Junxi Zhang, Xin Zhang, Qingzhao Ni, Yang Chen, Maoliang Wu and Chao Yang, Distinction in Corrosion Resistance of Selective Laser Melted Ti-6Al-4V Alloy on Different Planes, 703–710, 2016, with kind permission from Elsevier).

composition of phases in the planes may be possible due to different scan strategies used for the deposition. The results of the immersion test comply with the electrochemical experiments displaying that the XY plane lost weight amounting to 0.7 mg/cm$^2$ as compared to 0.9 mg/cm$^2$ in the XZ plane. The corrosion behaviour in harsher environments as well as the difference in corrosion resistance between the two planes are observed to be more pronounced.

EBM-built samples also show the variation in properties with the changes in direction of the planes under consideration. The change in corrosion behaviour with the change in planes is subsequently reported (Zhang and Qin 2019; Fojt et al. 2018).

The study of the corrosion behaviour of EBM-built Ti6Al4V along the planes at angles of 0°, 45°, 55° and 90° from the building direction was carried out by Gong et al. 2017 in a 1M HCl solution. Gas-atomised Ti-6Al-4V powders of the size of 40 to 120 µm are preheated at 843 K and melted using an electron gun with a beam size of 150 µm. The process is performed under a vacuum of $6.4 \times 10^{-3}$ mbar and at a voltage of 60 kV. The samples are exposed to 1M HCl coupled to a calomel reference electrode. The corrosion test was conducted five times at a scan rate of 1mV/s. The electrochemical results show that the $I_{corr}$ value descends in the order of 45°, 90°, 55° and 0°, which leads to the fact that the corrosion resistance of the planes reduces in also that order: 0°, 55°, 90° and 45°. The corrosion resistance is identified to be a function of the grain boundary per unit length, as shown in Figure 13.7. The quantity of α phase and β phase present in the surface also influences the corrosion resistance of the sample under consideration. Since EBM uses preheating, significant diffusion of constituent elements occurs in the samples, and the phase formation in EBM is diffusion controlled. As a result, the texture formation as noticed in the previous LPBF example does not happen in EBM-based samples (Dai et al. 2016). In this point of view, grain orientation does not affect the

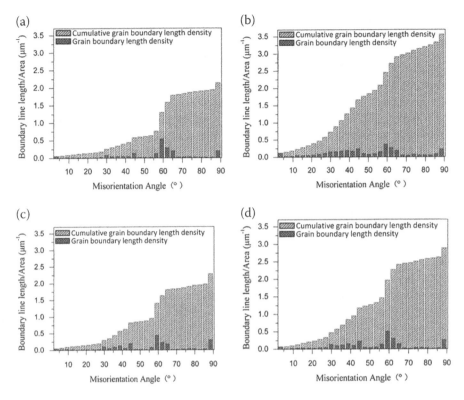

**FIGURE 13.7** Grain Boundary Length Density and Cumulative Rrain Boundary Length Density as a Function of Misorientation Angle at Planes (a) 0° (b) 45° (c) 55° and (d) 90° Sample (Reprinted from Corrosion Science, 127, Gong, Xiaojuan, Yujie Cui, Daixiu Wei, Bin Liu, Ruiping Liu, Yan Nie and Yunping Li, Building Direction Dependence of Corrosion Resistance Property of Ti–6Al–4V Alloy Fabricated by Electron Beam Melting, 101–109, 2017, with kind permission from Elsevier).

corrosion behaviour of the samples. The micrographs obtained after the corrosion study support the trend of the Icorr values.

As previously discussed, due to the effect of anisotropy, the properties of MAM-built samples can deviate from that of their wrought counterparts. The EBM-built samples are compared for their corrosion resistance with wrought counterparts by Abdeen and Palmer (2016) and Devika, Dass, and Chaudhary (2015).

Bai et al. (2017) compared the electrochemical corrosion behaviour of EBM-built Ti6Al4V with that of wrought Ti6Al4V alloy. Medical grade Ti6Al4V powder with a particle size of 51 μm to 109 μm is melted using a 60 kV electron beam with a beam size of 200 μm. The powders were deposited under vacuum at $10^{-3}$ mbar and were preheated to 1,003 K by electron beam prescanning. The wrought sample is manufactured by billet forging at 1,040°C. The electrochemical tests were run at a potential scan rate of 0.166 mV/s, with reference to a saturated calomel electrode in phosphate-buffered saline solution, at 37°C. The electrochemical results show a slight difference in Ecorr value for EBM-built Ti6Al4V ($-0.28 \pm 0.03$ V) and wrought Ti6Al4V (−0.29 ± 0.02 V). The Icorr values of the EBM-built Ti6Al4V (0.007 ± 0.001 μA·cm$^{-2}$) and wrought Ti6Al4V (0.012 ± 0.001 μA·cm$^{-2}$) are also rather near, with the wrought Ti6Al4V showing a slight superiority. The curves in Figure 13.8 reveal that the corrosion resistance of wrought Ti6Al4V is slightly inferior to EBM-built Ti6Al4V. This inferiority corresponds to the higher fraction of β phase, and the refined morphology of lamellar α/β phases in the microstructure of EBM-built Ti6Al4V. Coarse plate-shaped α phases with intergranular β phase are observed in the wrought sample, whereas the rod-shaped β phase and fine α platelets are found in the EBM-built sample due to higher cooling rate. The diffusion of α stabilising elements, such as Al, to α phase and β stabilising elements, like V, results in the difference in the composition of α and β plates and, moreover, to the generation of a potential difference at the grain boundary. However, the ultrafine-grained lamellar phases of α and β reduce element segregation at the grain boundaries and lower the galvanic effect. In addition to this, the ultrafine lamellar structure is believed to bear more passivation nucleation sites and lessens corrosion due to protective layer formation.

The high cooling rate of LPBF alters the microstructure of the samples from its conventional counterparts. Numerous studies are conducted in order to compare the corrosion resistance of LPBF- built Ti6Al4V with its conventional counterparts (Acquesta et al. 2019; Toptan et al. 2019).

Hemmasian Ettefagh et al. (2019) compared the corrosion behaviour of LPBF-manufactured Ti6Al4V to that of conventionally manufactured samples. Furthermore, the LPBF-built samples are heat-treated, and the change in corrosion behaviour is reported. The Ti6Al4V samples, which are manufactured with a laser power of 95–100 W and at a scan speed of 900–1000 mm/sec, are compared with cold-rolled conventional counterparts in 3.5 wt% NaCl at room temperature.

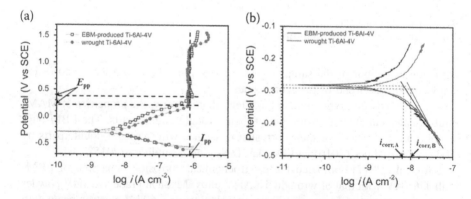

**FIGURE 13.8** (a) Potentiodynamic Curves of EBM-Built and Conventional Ti-6Al-4V alloy in Phosphate-Buffered Saline at 37 °C; (b) Tafel Extrapolation Method. (icorr, A) and (icorr, B) represent the corrosion current density EBM-produced and wrought Ti-6Al-4V alloy, respectively. (Reprinted from Corrosion Science, 123, Bai, Yun, Xin Gai, Shujun Li, Lai-Chang Zhang, Yujing Liu, Yulin Hao, Xing Zhang, Rui Yang and Yongbo Gao, Improved Corrosion Behaviour of Electron Beam Melted Ti-6Al–4V Alloy in Phosphate Buffered Saline, 289–296, 2017, with kind permission from Elsevier).

Additionally, the LPBF-built samples are annealed separately at temperatures of 800°C and 600°C for two hours in an argon environment. The samples were tested at a scan rate of 1.67 mV/s. The electrochemical test results reveal that the protective layer formed on the LPBF-built sample is less stable, and the corrosion rate is almost 16 times than that of the wrought sample. This is due to the unavailability of stable β phase as a result of the rapid solidification during LPBF. On heat treatment, the martensitic phase releases the stress point, and the BCC β phase forms. This results in increased corrosion resistance are shown in Figure 13.9. α and β phases in the microstructure are revealed in the SEM. Martensitic α′ phase is identified as a needle-shaped structure in the as-built LPBF samples. HT600 sample also shows a similar microstructure. On heat treating at 800°C, needle-shaped non-equilibrium phases become transformed to α and β phases that are comparable to the wrought sample. The microscopy of the corroded samples supports the corrosion behaviour in the samples.

Furthermore, WAAM-manufactured titanium alloys are compared to the samples built by other MAM technologies, as well as with conventional counterparts, for their corrosion behaviour (Yang et al. 2017; Ron et al. 2020).

The anisotropy in corrosion behaviour of WAAM-manufactured Ti6Al4V is compared to its conventional counterpart by Wu, Pan, Li et al. (2018). The samples are GTAW deposited at a deposition current, arc voltage, scan speed and wire feed speed of 110 A, 12 V, 95 mm/min and 1,000 mm/min, respectively. A 15-layered wall measuring 150 mm in length and 10 mm in width is deposited. The walls were sliced into vertical planes (VP) and horizontal planes (HP) and compared for corrosion behaviour with conventional cold-rolled samples that are sliced from the substrate base metal (BM) in 3.5% NaCl solution. The passive film resistance (Rf) of the BM samples (1.68 MΩ cm$^2$) is observed to be far greater compared to that of the

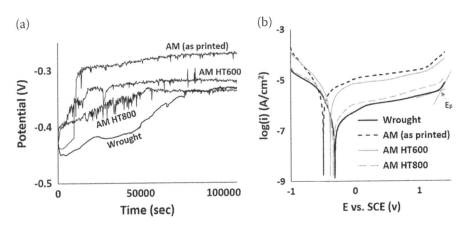

**FIGURE 13.9**    (a) Open Circuit Potential with Respect to Time, and (b) Tafel Curves of the Tested Samples. All tests were performed in 3.5 wt % NaCl water solution. (Reprinted from Additive Manufacturing, 28, Hemmasian Ettefagh, Ali, Congyuan Zeng, Shengmin Guo and Jonathan Raush, Corrosion Behaviour of Additively Manufactured Ti-6Al-4V Parts and the Effect of Post Annealing, 252–258, 2019, with kind permission from Elsevier).

(a)                          (b)              (c)

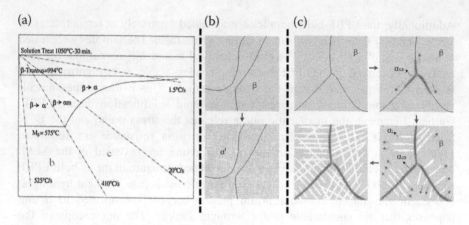

**FIGURE 13.10** Microstructural Evolution for WAAM-Built Ti-6Al-4V: (a) Continuous Cooling Diagram for Ti–6Al–4V β-Solution Treated at 1,050°C for 30 minutes, (b) HP regions, (c) VP regions. (Reprinted from Corrosion Science, 137, Wu, Bintao, Zengxi Pan, Siyuan Li, Dominic Cuiuri, Donghong Ding and Huijun Li, The Anisotropic Corrosion Behaviour of Wire Arc Additive Manufactured Ti-6Al-4V Alloy in 3.5% NaCl Solution, 176–183, 2018, with kind permission from Elsevier).

VP (195 kΩ cm$^2$) and HP (174 kΩ cm$^2$) samples, which points to the fact that the corrosion resistance diminishes in the order BM > VP > HP. The change in the corrosion rate in VP and HP samples is due to the microstructural changes, as presented in Figure 13.10. Coarse lamellar α grains interwoven with Widmanstätten structure are reported to be formed in the VP due to the slow cooling rate. The faster cooling rate along the HP lead to the formation of α' grains. The lower dissolution of α grains and the availability of more passivation sites make VP more corrosion resistant. β phases are present only in BM samples and correspond to the higher corrosion resistance than the WAAM-built sample. and the uniformity of microstructure in BM samples resulted in the repeatability of the corrosion tests, as compared to VP and HP samples.

## 13.4 CONCLUSIONS

The electrochemical corrosion behaviour of MAM-built alloys has been reviewed. The change in process and parameters modifies the grain morphology and grain size and consequently affects the corrosion behaviour of the metals. The non-uniform microstructure of MAM samples reduces the repeatability of corrosion testing. Large anisotropy is observed in corrosion behaviour of MAM-built components. Corrosion is identified to initiate at pores. During heat treatment, the anisotropy of MAM parts is diminished, and the directionality of corrosion properties reduces. Furthermore, the change in corrosion resistance amidst heat treatment depends on the material and the treating temperature. Element precipitation, phase formation, element segregation, etc. affects corrosion resistance. In order to control the corrosion of MAM samples, more studies regarding the mechanism of corrosion and the influence of metallurgy on the corrosion must be carried out.

## ACKNOWLEDGEMENTS

Mr. P.K. Diljith and Mr. A.N. Jinoop acknowledge the financial support from the Ministry of Human Resources Development and Department of Atomic Energy, Government of India, respectively. The authors thank the members of the LAM lab, RRCAT for their help during the study.

## REFERENCES

Abdeen, Dana H., and Bruce R. Palmer. 2016. "Corrosion Evaluation of Ti-6Al-4V Parts Produced with Electron Beam Melting Machine." *Rapid Prototyping Journal* 22 (2). Emerald Group Publishing Limited: 322–329. doi:10.1108/RPJ-09-2014-01 04.

Acquesta, Annalisa, Anna Carangelo, Paolo Di Petta, and Tullio Monetta. 2019. "Electrochemical Characterization of Ti6Al4V Components Produced by Additive Manufacturing." *Key Engineering Materials* 813. Trans Tech Publications Ltd: 86–91. doi:10.4028/www.scientific.net/KEM.813.86.

Bai, Yun, Xin Gai, Shujun Li, Lai-Chang Zhang, Yujing Liu, Yulin Hao, Xing Zhang, Rui Yang, and Yongbo Gao. 2017. "Improved Corrosion Behaviour of Electron Beam Melted Ti-6Al–4V Alloy in Phosphate Buffered Saline." *Corrosion Science* 123: 289–296. doi:https://doi.org/10.1016/j.corsci.2017.05.003.

Buchanan, R. A., and E. E. Stansbury. 2012. "4 - Electrochemical Corrosion." In, edited by B. T. Myer, *Handbook of Environmental Degradation of Materials (Second Edition) Kutz*, 87–125. Oxford: William Andrew Publishing. doi:https://doi.org/10.1016/B978-1-4377-3455-3.00004-3.

Chandramohan, P., Shepherd Bhero, Babatunde Abiodun Obadele, and Peter Apata Olubambi. 2017. "Laser Additive Manufactured Ti–6Al–4V Alloy: Tribology and Corrosion Studies." *The International Journal of Advanced Manufacturing Technology* 92 (5): 3051–3061. doi:10.1007/s00170-017-0410-2.

Chen, L. Y., J. C. Huang, C. H. Lin, C. T. Pan, S. Y. Chen, T. L. Yang, D. Y. Lin, H. K. Lin, and J. S. C. Jang. 2017. "Anisotropic Response of Ti-6Al-4V Alloy Fabricated by 3D Printing Selective Laser Melting." *Materials Science and Engineering: A* 682: 389–395. doi:https://doi.org/10.1016/j.msea.2016.11.061.

Chiu, Tse-Ming, Mohamad Mahmoudi, Wei Dai, Alaa Elwany, Hong Liang, and Homero Castaneda. 2018. "Corrosion Assessment of Ti-6Al-4V Fabricated Using Laser Powder-Bed Fusion Additive Manufacturing." *Electrochimica Acta* 279: 143–151. doi:https://doi.org/10.1016/j.electacta.2018.04.189.

Criales, Luis E., Yiğit M. Arısoy, Brandon Lane, Shawn Moylan, Alkan Donmez, and Tuğrul Özel. 2017. "Laser Powder Bed Fusion of Nickel Alloy 625: Experimental Investigations of Effects of Process Parameters on Melt Pool Size and Shape with Spatter Analysis." *International Journal of Machine Tools and Manufacture* 121: 22–36. doi:https://doi.org/10.1016/j.ijmachtools.2017.03.004.

Dai, Nianwei, Lai-Chang Zhang, Junxi Zhang, Xin Zhang, Qingzhao Ni, Yang Chen, Maoliang Wu, and Chao Yang. 2016. "Distinction in Corrosion Resistance of Selective Laser Melted Ti-6Al-4V Alloy on Different Planes." *Corrosion Science* 111: 703–710. doi:https://doi.org/10.1016/j.corsci.2016.06.009.

Devika, D, Soumya Shekhar Dass, and Suneel Kumar Chaudhary. 2015. "Characterization and Corrosion Behaviour Study on Biocompatible Ti-6Al-4V Component Fabricated by Electron Beam Melting." *Journal of Biomimetics, Biomaterials and Biomedical Engineering* 22. Trans Tech Publications Ltd: 63–75. doi:10.4028/www.scientific.net/JBBBE.22.63.

Fedorov, Vasiliy, Vasiliy Klimenov, Roman Cherepanov, and Andrey Batranin. 2019. "Powder and Wire Melting of Titanium Alloys by Electron Beam." *Procedia Manufacturing* 37: 584–591. doi:https://doi.org/10.1016/j.promfg.2019.12.092.

Fojt, Jaroslav, Michaela Fousova, Eva Jablonska, Ludek Joska, Vojtech Hybasek, Eva Pruchova, Dalibor Vojtech, and Tomas Ruml. 2018. "Corrosion Behaviour and Cell Interaction of Ti-6Al-4V Alloy Prepared by Two Techniques of 3D Printing." *Materials Science and Engineering: C* 93: 911–920. doi:https://doi.org/10.1016/j.msec.2018.08.066.

Gong, Xiaojuan, Yujie Cui, Daixiu Wei, Bin Liu, Ruiping Liu, Yan Nie, and Yunping Li. 2017. "Building Direction Dependence of Corrosion Resistance Property of Ti–6Al–4V Alloy Fabricated by Electron Beam Melting." *Corrosion Science* 127: 101–109. doi:https://doi.org/10.1016/j.corsci.2017.08.008.

Graf, Benjamin, Stefan Ammer, Andrey Gumenyuk, and Michael Rethmeier. 2013. "Design of Experiments for Laser Metal Deposition in Maintenance, Repair and Overhaul Applications." *Procedia CIRP* 11: 245–248. doi:https://doi.org/10.1016/j.procir.2013.07.031.

Guo, Pengfei, Xin Lin, Jiaqiang Li, Yufeng Zhang, Menghua Song, and Weidong Huang. 2018. "Electrochemical Behavior of Inconel 718 Fabricated by Laser Solid Forming on Different Sections." *Corrosion Science* 132 (September 2017). Elsevier: 79–89. doi:1 0.1016/j.corsci.2017.12.021.

Hemmasian Ettefagh, Ali, Congyuan Zeng, Shengmin Guo, and Jonathan Raush. 2019. "Corrosion Behavior of Additively Manufactured Ti-6Al-4V Parts and the Effect of Post Annealing." *Additive Manufacturing* 28: 252–258. doi:https://doi.org/10.1016/j.addma.2019.05.011.

Jinoop, A. N., C. P. Paul, and K. S. Bindra. 2019. "Laser-Assisted Directed Energy Deposition of Nickel Super Alloys: A Review." *Proceedings of the Institution of Mechanical Engineers, Part L: Journal of Materials: Design and Applications* 233 (11): 2376–2400. doi:10.1177/1464420719852658.

Jinoop, A. N., C. P. Paul, S. K. Mishra, and K. S. Bindra. 2019. "Laser Additive Manufacturing Using Directed Energy Deposition of Inconel-718 Wall Structures with Tailored Characteristics." *Vacuum*166 (March). Elsevier: 270–278. doi:10.1016/j.vacuum.2019.05.027.

Kong, Decheng, Chaofang Dong, Xiaoqing Ni, Liang Zhang, Cheng Man, Guoliang Zhu, Jizheng Yao, Li Wang, Xuequn Cheng, and Xiaogang Li. 2019. "Effect of TiC Content on the Mechanical and Corrosion Properties of Inconel 718 Alloy Fabricated by a High-Throughput Dual-Feed Laser Metal Deposition System." *Journal of Alloys and Compounds* 803. Elsevier B.V: 637–648. doi:10.1016/j.jallcom.2019.06.317.

Kurzynowski, Tomasz, Irina Smolina, Karol Kobiela, Bogumiła Kuźnicka, and Edward Chlebus. 2017. "Wear and Corrosion Behaviour of Inconel 718 Laser Surface Alloyed with Rhenium." *Materials & Design* 132: 349–359. doi:https://doi.org/10.1016/j.matdes.2017.07.024.

Li, Jiaqiang, Xin Lin, Jian Wang, Min Zheng, Pengfei Guo, Yufeng Zhang, Yongming Ren, Jianrui Liu, and Weidong Huang. 2019. "Effect of Stress-Relief Annealing on Anodic Dissolution Behaviour of Additive Manufactured Ti-6Al-4V via Laser Solid Forming." *Corrosion Science* 153: 314–326. doi:https://doi.org/10.1016/j.corsci.2019.04.002.

Li, Johnnieew, Mohd Rizal Alkahari, Nor Rosli, Rafidah Hasan, M N Sudin, and Faiz Ramli. 2019. "Review of Wire Arc Additive Manufacturing for 3D Metal Printing." *International Journal of Automation Technology* 13 (May): 346–353. doi:10.20965/ijat.2019.p0346.

Luo, Shuncun, Wenpu Huang, Huihui Yang, Jingjing Yang, Zemin Wang, and Xiaoyan Zeng. 2019. "Microstructural Evolution and Corrosion Behaviors of Inconel 718 Alloy Produced by Selective Laser Melting Following Different Heat Treatments." *Additive Manufacturing* 30: 100875. doi:https://doi.org/10.1016/j.addma.2019.100875.

Mahamood, R. M., and E. T. Akinlabi. 2018. "Corrosion Behavior of Laser Additive Manufactured Titanium Alloy." *The International Journal of Advanced Manufacturing Technology* 99 (5): 1545–1552. doi:10.1007/s00170-018-2537-1.

Majchrowicz, Kamil, Zbigniew Pakieła, Janusz Kamiński, Magdalena Płocińska, Tomasz Kurzynowski, and Edward Chlebus. 2018. "The Effect of Rhenium Addition on Microstructure and Corrosion Resistance of Inconel 718 Processed by Selective Laser Melting." *Metallurgical and Materials Transactions A* 49 (12): 6479–6489. doi:1 0.1007/s11661-018-4926-3.

Nayak, S. K., S. K. Mishra, C. P. Paul, A. N. Jinoop, and K. S. Bindra. 2020. "Effect of Energy Density on Laser Powder Bed Fusion Built Single Tracks and Thin Wall Structures with 100 Mm Preplaced Powder Layer Thickness." *Optics & Laser Technology* 125: 106016. doi:https://doi.org/10.1016/j.optlastec.2019.106016.

Popovich, V. A., E. V. Borisov, A. A. Popovich, V. Sh. Sufiiarov, D. V. Masaylo, and L. Alzina. 2017. "Impact of Heat Treatment on Mechanical Behaviour of Inconel 718 Processed with Tailored Microstructure by Selective Laser Melting." *Materials & Design* 131: 12–22. doi:https://doi.org/10.1016/j.matdes.2017.05.065.

Raj, B. Anush, J. T. Winowlin Jappes, M. Adam Khan, V. Dillibabu, and N. C. Brintha. 2019. "Studies on Heat Treatment and Electrochemical Behaviour of 3D Printed DMLS Processed Nickel-Based Superalloy." *Applied Physics A* 125 (10): 722. doi:10.1 007/s00339-019-3019-5.

Rao, V. Shankar. 2004. "A Review of the Electrochemical Corrosion Behaviour of Iron Aluminides." *Electrochimica Acta* 49 (26): 4533–4542. doi:https://doi.org/10.1016/ j.electacta.2004.05.033.

Ron, Tomer, Galit Katarivas Levy, Ohad Dolev, Avi Leon, Amnon Shirizly, and Eli Aghion. 2020. "The Effect of Microstructural Imperfections on Corrosion Fatigue of Additively Manufactured ER70S-6 Alloy Produced by Wire Arc Deposition." *Metals* 10 (1). Multidisciplinary Digital Publishing Institute: 98.

Shen, Chen, Gang Mu, Xueming Hua, Fang Li, Dongzhi Luo, Xiangru Ji, and Chi Zhang. 2019. "Influences of Postproduction Heat Treatments on the Material Anisotropy of Nickel-Aluminum Bronze Fabricated Using Wire-Arc Additive Manufacturing Process." *The International Journal of Advanced Manufacturing Technology* 103 (5): 3199–3209. doi:10.1007/s00170-019-03700-7.

Shen, Chen, Zengxi Pan, Donghong Ding, Lei Yuan, Ning Nie, Ying Wang, Dongzhi Luo, Dominic Cuiuri, Stephen van Duin, and Huijun Li. 2018. "The Influence of Post-Production Heat Treatment on the Multi-Directional Properties of Nickel-Aluminum Bronze Alloy Fabricated Using Wire-Arc Additive Manufacturing Process." *Additive Manufacturing* 23: 411–421. doi:https://doi.org/10.1016/ j.addma.2018.08.008.

Thirumalaikumarasamy, D., K. Shanmugam, and V. Balasubramanian. 2014. "Comparison of the Corrosion Behaviour of AZ31B Magnesium Alloy under Immersion Test and Potentiodynamic Polarization Test in NaCl Solution." *Journal of Magnesium and Alloys* 2 (1): 36–49. doi:https://doi.org/10.1016/j.jma.2014.01.004.

Toptan, Fatih, Alexandra C. Alves, Óscar Carvalho, Flávio Bartolomeu, Ana M. P. Pinto, Filipe Silva, and Georgina Miranda. 2019. "Corrosion and Tribocorrosion Behaviour of Ti6Al4V Produced by Selective Laser Melting and Hot Pressing in Comparison with the Commercial Alloy." *Journal of Materials Processing Technology* 266: 239–245. doi:https://doi.org/10.1016/j.jmatprotec.2018.11.008.

Wu, Bintao, Zengxi Pan, Donghong Ding, Dominic Cuiuri, Huijun Li, Jing Xu, and John Norrish. 2018. "A Review of the Wire Arc Additive Manufacturing of Metals: Properties, Defects and Quality Improvement." *Journal of Manufacturing Processes* 35: 127–139. doi:https://doi.org/10.1016/j.jmapro.2018.08.001.

Wu, Bintao, Zengxi Pan, Siyuan Li, Dominic Cuiuri, Donghong Ding, and Huijun Li. 2018. "The Anisotropic Corrosion Behaviour of Wire Arc Additive Manufactured Ti-6Al-4V Alloy in 3.5% NaCl Solution." *Corrosion Science* 137: 176–183. doi:https://doi.org/1 0.1016/j.corsci.2018.03.047.

Xu, Yangzi, Yuan Lu, Kristin L. Sundberg, Jianyu Liang, and Richard D. Sisson. 2017. "Effect of Annealing Treatments on the Microstructure, Mechanical Properties and Corrosion Behavior of Direct Metal Laser Sintered Ti-6Al-4V." *Journal of Materials Engineering and Performance* 26 (6): 2572–2582. doi:10.1007/s11665-01 7-2710-y.

Yadav, Sunil, Christ P. Paul, Arackal N. Jinoop, Saurav K. Nayak, Arun K. Rai, and Kushvinder S. Bindra. 2019. "Effect of Process Parameters on Laser Directed Energy Deposition of Copper." ASME 2019 Gas Turbine India Conference. doi:10.1115/ GTINDIA2019-2453.

Yakout, Mostafa, M. A. Elbestawi, and Stephen C. Veldhuis. 2018. "A Review of Metal Additive Manufacturing Technologies." *Solid State Phenomena* 278. Trans Tech Publications Ltd: 1–14. doi:10.4028/www.scientific.net/SSP.278.1.

Yang, Jingjing, Huihui Yang, Hanchen Yu, Zemin Wang, and Xiaoyan Zeng. 2017. "Corrosion Behavior of Additive Manufactured Ti-6Al-4V Alloy in NaCl Solution." *Metallurgical and Materials Transactions A* 48 (7): 3583–3593. doi:10.1007/s11661- 017-4087-9.

Zhang, L. N., and O. A. Ojo. 2020. "Corrosion Behavior of Wire Arc Additive Manufactured Inconel 718 Superalloy." *Journal of Alloys and Compounds* 829: 154455. doi:https:// doi.org/10.1016/j.jallcom.2020.154455.

Zhang, Lai-Chang, and Peng Qin. 2019. "Corrosion Behaviors of Additive Manufactured Titanium Alloys." In *Additive Manufacturing of Emerging Materials*, edited by Bandar AlMangour, 197–226. Cham: Springer International Publishing. doi:10.1007/978-3-31 9-91713-9_6.

Zhang, Yufeng, Xin Lin, Jun Yu, Pengfei Guo, Jiaqiang Li, Tuo Qin, Jianrui Liu, and Weidong Huang. 2020. "Electrochemical Dissolution Behavior of Heat Treated Laser Solid Formed Inconel718." *Corrosion Science* 108750. doi:https://doi.org/10.1016/ j.corsci.2020.108750.

# 14 A Review of Nonlinear Control for Electrohydraulic Actuator System in Automation

*Ashok Kumar Kumawat, Manish Rawat,
Renu Kumawat, and Raja Rout*

## CONTENTS

## 14.1 INTRODUCTION

The hydraulic system has become rather popular in modern industries due to its ability to handle large inertia and heavyweight loads. Technological advancements in the hydraulic actuation system provide high power to size ratio with swift response and high accuracy compared with electrical drive systems. In spite of these advantages, some challenges, such as valve nonlinearity, friction nonlinearity of the actuator, parametric uncertainties, unmodeled dynamics and unknown external disturbances in EHAS are still significant topics in industry and academic research. Various application areas are: wind power transfer systems (Vaezi and Izadian, 2015), power steering systems (Dell'Amico and Krus, 2015), vehicle suspension systems (Ekoru and Pedro, 2013), autonomous excavators (Sotiropoulos and Asada, 2019) structural testing (Shen et al., 2016), exoskeletons (Chen et al., 2018), hydraulic turbines (Chen et al., 2015), aircraft actuation systems (Ijaz et al., 2018) and robotic manipulators (Asl et al., 2020). For adequate position and velocity tracking of linear and rotational motion, electro-hydraulic valves, such as the proportional valve and the servo valve, were used in the literature. The servo valve provides more accurate position control rather than the proportional valve; therefore, in most of the literature, the use of servo in hydraulic system control is mentioned and discussed. In Wang et al. (2014a), the spool dynamics of the servo valve were omitted to make the complete system modelling less complex. For example, uncertainty due to viscous friction on the spool, variation in oil temperature and valve saturation conditions were presented by Guo et al. (2016). These uncertainties were also considered time-dependent and can create instability in position tracking (Yao et al., 2014). In Wang et al. (2017), the fourth-order Butterworth filter was implemented in order to attenuate the measurement of noise in the pressure and position parameter. For the management of linear systems, the proportional–integral–derivative (PID) controller is a rather useful technique, but it fails under nonlinear conditions. Therefore, PID control cannot be used for highly accurate control of EHAS. In Wang et al. 2014a, 2014b, the Lyapunov analysis was utilised for nonlinear system stability. In addition, a nonlinear adaptive controller was presented using the backstepping approach to be able to achieve good tracking performance while considering internal leakage flow. In regard to further improvement for the adaptive controller, ESO-based control algorithm was implemented in order to achieve considerable tracking performance (Wang et al., 2017). However, position error was increased due to an increment in feedback gain. In Wang et al., 2015a, 2015b, motion disturbance due to the actuator was suppressed by applying robust control which consists of two loops, such as the open loop and the closed loop, to increase the robustness of the proposed controller. However, it was assumed that disturbances were differentiable and bound. This chapter presents an expanded literature review on nonlinear control schemes for EHAS with diverse challenges under various disturbances and uncertainties by examining existing literature from 2011 to 2020. The chapter is organised as follows: various types of challenges in modelling are presented in Section 14.2. In Section 14.3, assorted control schemes such as linear and nonlinear control are discussed. Finally, Section 14.4 concludes the chapter.

## 14.2 REVIEW OF SYSTEM MODELLING

Automation in industries is rapidly increasing due to development in technology. In hydraulic systems, electrohydraulic valves play an extremely important role in heavy industries. There is a need for highly accurate and precise position tracking for the heavy payload in rectilinear and rotational motion-based applications. For better understanding and management of the EHAS system, modelling is required. Therefore, a control signal can be defined in terms of other parameters and system states. In addition, some states and parameters might be unknown and uncertain which makes modelling complex. These nonlinearity/parametric uncertainties can be estimated or neglected, as per system stability.

### 14.2.1 CHALLENGES IN EHAS MODELLING

EHAS is an electromechanical system that is controlled by electrohydraulic valves, such as the servo valve and the proportional valve. The complete mathematical model for an EHAS can be defined in a linear and nonlinear mode, as per the accuracy of the actuator output. However, certain challenges appear in EHAS implementation; an example of which is the modelling of a complete system in a nonlinear manner and the integration of the mechanical system with electrical and electronics systems. Generally, uncertainties were classified into two types: Structured and unstructured. Uncertainty due to internal leakage is considered in designing the control law for the nonlinear hydraulic system, as per Ahn et al. (2013). However, experimentally, some parameters were estimated within the satisfactory limit for good tracking response. Sometimes, variations in supply pressure may possibly be caused by parameter uncertainty (Mintsa et al. (2011). In addition, measurement noise and servo valve saturation nonlinearity were discussed and consequently added poor output performance in the case of the PID controller. Busquets and Ivantysynova (2015) withheld the dynamics of the spool in checking the valve to reduce the complexity in modelling of EHAS. In the steering system, instability was produced by a large wheel torque and fixed gain-bandwidth product, which was discussed in Dell'Amico and Krus (2015). In addition, the first-order filter was used to improve stability in the pressure loop, and the PI controller was used. Guo et al. (2016) discussed the variation in viscous friction, oil temperature and valve saturation characteristics, which were the cause of parametric uncertainties in EHAS. In addition, nominal values of these parameters were considered in state estimation. In the field of active suspension systems, external disturbances were also considered, such as road disturbance, which was presented in discrete waveform nature with a small amplitude (Sun et al., 2012). In wind power transfer systems, a greater number of hydraulic units increased the nonlinear complexity in mathematical modelling, because, in the hydraulic, each component bears various types of nonlinearity as well as parametric uncertainty. However, the system was approximately linearised, accuracy was limited to around 91% (Vaezi and Izadian, 2015). In Wang et al. (2015a) and servo valve dynamics were neglected due to the fast response of spool rather than test rig, and it was considered linear relation with a control input. In addition, uncertainty terms were assumed to be bound, along with its derivative terms. However, derivative gain increased the possibility of measurement noise in torque and poor system performance, so it was considered as zero in order to design the PID controller.

### 14.2.2 DIFFERENT MODELLING APPROACHES

In the existing literature, various modelling techniques have been analysed. In this chapter, we focus mainly on three types of modelling approaches for an EHAS system. The mathematical modelling of EHAS is quite important in order to define the control variable and to estimate unknown parameters and states. A detailed discussion on the modelling approach for EHAS is presented in the further subsections, and a summary is shown in Table 14.1.

### 14.2.2.1   Model-Based

The model-based technique is widely used for the electromechanical system. In this chapter, most articles that used this approach are in the nonlinear model, and all states could be derived. A reference model was used to control the nonlinear robotic end-effector for variable payload (Zhang and Wei, 2017) and to estimate the unknown parameters. In addition, nonlinear terms were compensated by the adaptive control technique which improved the output tracking response. In Ye et al., 2020, a model-based technique was implemented on a pump-controlled electrohydraulic actuator which was had a saturation-type nonlinearity in the input as well as uncertainty in the friction model. To compensate for these issues, an adaptive commander filtering control was constructed using the model-based technique, and friction compensation was proposed. Furthermore, the complexity of management and online computation time was discovered as less than the classical backstepping approach. Further, the LuGre model was used as a friction compensator, which subsequently improved tracking precision. A model-based reference approach was used for the robotic manipulator and inverted pendulum in order to enhance the stability of the controller (Kumar et al., 2017). Di Massa et al. (2013) presented a model reference adaptive approach for seismic isolators in the presence of

### TABLE 14.1
### Research Category Based on Modelling Approach

| Research Category | Literature | Remarks |
|---|---|---|
| model-based | Di Massa et al. (2013); Yao et al. (2015a); Kumar et al. (2017); Zhang and Wei (2017); Ye et al. (2020) | • All system states are defined.<br>• A reference model can be used to define an unknown model<br>• The procedure is easy to implement. |
| system identification | Todeschini et al. (2014); Strano and Terzo (2016); Sui and Tong (2016); Pasolli and Ruderman (2018); Dang et al. (2019); Dong et al. (2019); Lu et al. (2019); Asl et al. (2020) | • Unknown states can be estimated.<br>• The filtering method can be used.<br>• Uncertainties and disturbances make system modelling more complex. |
| practical unknown | Wang et al. (2014a, 2014b); Yin and Zhao (2019); Kang et al. (2018); Xudong et al. (2018); Vaezi and Izadian (2015) | • Unknown system states and parameters can be estimated using a practical approach.<br>• An approximate model is estimated.<br>• Measurement noise can affect the system model. |

uncertainties that were created in operating conditions. In Yao et al., 2015a LuGre model-based friction compensation was implemented, and it established satisfactory tracking accuracy in the presence of the proposed controller. Based on the cases above, a model-based approach is a suitable approach to factor unknown model parameters and system states that help in achieving asymptotically tracking performance in EHAS.

### 14.2.2.2  System Identification

In many industrial applications, some parameters and states, which can be time-varying or invariant in EHAS, are unknown and uncertain. This kind of situation causes complexity in the modelling of the system, as well as in terms of its management. To cope with this problem, certain system identification tools can be used to formulate an approximate nonlinear model. Ahn et al. (2014) assumed that no external leakage with the four states was known, and unknown parameters were estimated to develop better system accuracy. In some applications, filter techniques were used for purposes of estimation; an example is when Asl et al. (2020) proposed a filter that was used to measure the unknown states of the robotic manipulator and servo-hydraulic systems while considering time-dependent noise. However, the proposed filter technique was deemed inappropriate for multiobjective optimisation problems. In Dang et al. (2019), force and pressure model dynamics were estimated by applying a disturbance observer in order to achieve better output precision. In Dong et al. (2019), an adaptive parameter estimator was adopted to achieve satisfactory tracking performance, as well as the stability of the controller which included Levant's differentiator and adaptive law. Pasolli and Ruderman (2018) highlighted modelling and identification with the consideration of a full order, linearised and reduced model. Another nonlinear estimation approach in the Kalman filter, which was based on the state-dependent Riccati-Equation, was implemented on a fifth-order nonlinear system (Strano and Terzo, 2016). Moreover, simulated and experimental results showed the applicability of the proposed filter by attenuating measurement noise and hard nonlinearities. Sui and Tong (2016) discussed the switched fuzzy state observer to estimating the immeasurable state. The proposed observer enhanced the stability of the control design problem which was verified using the Lyapunov function and average dwell time technique. However, the usability of adaptive fuzzy control for multi-input, multi-output or large-scale switched nonlinear systems was left for further discussion. Lu et al. (2019) proposed a novel adaptive unscented Kalman filter to estimate the disturbances in real-time, with the existence of sensor noise. Model identification was done and high robustness was noted in the presence of cascade control (Todeschini et al., 2014).

### 14.2.2.3  Practical Unknown

This approach is the third one used to develop the model of an EHAS. In this procedure, unknown parameters and states can be defined by using a practical approach that is based on simulation or experimental implementation. In Kang et al. (2018), the GMS friction model was identified experimentally and divided into two steps. The first step was the identification of static friction parameters, and the second step was determining dynamic parameters such as weight and stiffness coefficients. Hydraulic-based wind power transfer systems are widely used as renewable energy systems, but the complete systems exhibited nonlinear terms due to the presence of actuator hysteresis, disturbances in load torque and wind speed (Vaezi and Izadian, 2015). In addition, a graphical approach used to establish the average nonlinear model and sub-models were also derived using experimental analysis. The simulated results were almost identical with the experimental data. Flow nonlinearity, motion disturbance and friction model uncertainty were discussed for the electro-hydraulic load simulators (Wang et al., 2014a). The friction model was compensated using an experimentally identified model which increases the stability of the system based on a single-step backstepping design process. Further, Lyapunov criteria verified the stability of the output response. In Xudong et al. (2018), a nonlinear servo-hydraulic system shows instability in the presence of mismatched disturbances, which was compensated by the estimation process. Furthermore, a close-loop controller was implemented in order to minimise tracking error. Experimentally, friction data was defined, and stability was verified of EHS (Wang et al., 2014b). Yin and Zhao (2019) presented the modelling of the hydrostatic tidal turbine with parametric uncertainties and nonlinearities with hydrostatic dynamics. A nonlinear observer using an extreme learning machine was developed to achieve speed and torque tracking. Moreover, based on the observed parameters, sensorless double integral SMC was implemented under the designed modelling conditions. However, optimal power control with sensorless scheme deployment on other test rigs remains open for future work in terms of the hardware loop mode.

## 14.3   REVIEW OF CONTROL STRATEGIES

It is particularly important to design a control algorithm under conditions of nonlinearity and uncertainties to attain highly accurate and precise tracking performance. In a hydraulic system, cylinders and motors are used to provide motion to an external load. In addition, the hydraulic supply to the cylinder and motor is controlled by electrohydraulic valves, such as the proportional valve and servo valve. As per the literature survey, the servo valve provides better position tracking performance rather than the proportional valve in terms of controlling the hydraulic motor and cylinder. However, the dynamics of the servo valve are more nonlinear and complex rather than the proportional valve. In EHAS, various internal and external disturbances occur, which brings about complexity in a system modelling and affects the tracking response. To cope with this condition, a nonlinear control scheme is very useful. The general nonlinear design approach for an EHAS is presented in Figure 14.1. In addition, if the EHAS system can be formulated in linear models, then the complexity of the system lessens, and the linear control algorithm can be implemented such as the PI, PD and PID controllers. However, there are also some challenges related to the controller gain parameter tuning and other modelling uncertainties. In order to compensate for this kind of problem, linear control can be implemented with a few optimisation techniques or a filtering process. In a nonlinear model, parametric variations and nonlinearities can be compensated by various

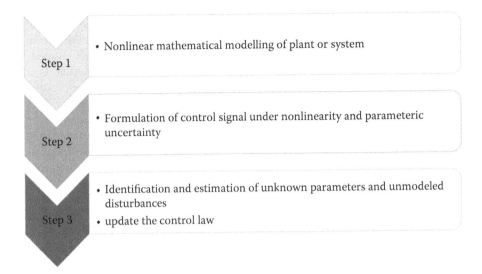

**FIGURE 14.1**   Design Steps of Nonlinear Control for EHAS.

types of nonlinear controllers, such as adaptive control, backstepping control, sliding mode control, ESO-based control and robust control, among others. Furthermore, the techniques and system stability can be analysed using Lyapunov criteria. If there is some measurement noise present in sensor output, then the measured signal can be passed through the designed filter, which improves tracking response and controller stability. In nonlinear EHAS, some states might be unmeasurable and unknown, but these states can be estimated by using a different kind of observation method, such as ESO. Moreover, to improve the stability of controller, robustness can be enhanced by implementing the robust control technique. Further classification of nonlinear and linear control schemes is discussed below.

## 14.3.1  LINEAR CONTROL TECHNIQUE

The proportional-integral-derivative (PID) controller was widely used as a linear controller in the literature for the EHAS. Using this, system modelling is done in linear approach, and input-output relation is defined in transfer function form. A feedback linearization concept was implemented on switching the position controller in the presence of pressure disturbances, and this uncovered better results than the classical PID controller (Mintsa et al., 2011). Further robustness was verified towards the measurement noise and nonlinearity in the actuator, but a variation in torque and friction were not considered. Chen et al. (2015) presented a linear model of hydraulic turbine system by considering negligible variations in nonlinear functions. Furthermore, the turbine system was controlled by the PID controller, and adaptive grid particle swarm optimisation was used to enhance the controller gains. The simulation results showed the robustness of the controller. In

Dell'Amico and Krus (2015), the linear model was analysed for open and closed-centre systems which were managed by a PI controller where boost gain depended on the bandwidth in the case of an open-centre system, as opposed to circumstances present in closed-centre systems. In addition, a faster valve was suggested in order to compensate for the extra phase shift due to the prefilter used to improve pressure tracking. Cascade control is also an advisable technique in handling linear EHAS models. For a nonlinear active vehicle suspension system, an inner PID control loop was used for force control, while an outer PID control loop was used for suspension control (Ekoru and Pedro, 2013). Moreover, the robustness of the system was verified via simulation results in the time and frequency domains with parametric uncertainty. Shen et al. (2016) discussed feedforward force controller, which included force inverse models and velocity compensators for force tracking performance under disturbances. Moreover, the mean position and root mean square error accuracy was improved by the proposed controller rather than the PID controller. The hydraulic excavator yielding dead zone and saturation-type nonlinearities were controlled by the PID controller, and its gain was tuned by using particle swarm optimisation to improve system control stability (Ye et al., 2017). Therefore, various limitations in linear control approaches for the EHAS system were due to modelled and unmodeled disturbances.

## 14.3.2 Nonlinear Control Technique

In practice, every electro-hydraulic actuator system bears structured and unstructured nonlinearities. Therefore, under these nonlinearities, nonlinear control algorithms are considered more suitable because of better system stability. A novel backstepping controller based on the bond graph model was proposed for a single-wheel brake system (Zhao et al., 2020). However, there is room for improvement with regards to developing a more accurate model for motor pumps and high-pressure accumulators for the given brake system. Zhao et al. (2019) discussed the hydraulic parallel manipulator which was managed by robust control using the backstepping technique. In addition, the cross-coupling control approach and joint disturbance rejection technique improved the stability of the system under model uncertainties and nonlinear coupling dynamics.

Dang et al. (2019) presented the nonlinear controller, which was a combination of the backstepping and sliding mode controller. However, the direction of the actuator and large bandwidth made the system unstable. Additionally, automatic tuning of controller gain was missing in the proposed controller. The quarter car suspension system initially had a modelling error which was removed by applying the Quasi-Newton algorithm over the complex space with iterative learning control (Wang et al., 2019). A three-layer adaptive control was used to handle the underwater parallel platform with complex dynamics and unknown disturbance (Xia et al., 2018). In Won et al. (2014), a disturbance observer with a high gain was used to estimate the load force, friction, measurement noise and parametric uncertainties. Moreover, the compensation of disturbances improved backstepping controller stability for position tracking errors. The process in improving transient behaviour, as discussed by Busquets and Iyantysynova

(2015), and the adaptive robust controller were implemented on a test bench with unmodeled uncertainties. However, spool dynamics of the valve were omitted. Cerman and Hušek (2012) proposed a fuzzy-based adaptive sliding mode controller to lessen the chattering noise found in the classical sliding mode controller, and tracking error was reduced for the hydraulic servo system. In Ijaz et al., 2018, settling time, percentage overshoot, steady-state error and force fighting reduction parameters were enhanced by using the adaptive integral sliding mode control technique, which was based on the polytopic linear parameter varying model. The fuzzy neural network-based sliding mode controller exhibited better performance for the hydraulic servo system in terms of settling time and steady-state error (Kayacan and Kaynak, 2012). Chen et al. (2018) applied the proposed adaptive robust controller for leg exoskeleton with 3-DOF under disturbances and load variations. Li and He (2015) proposed a nonlinear cascade control for hydraulic braking and disturbance. An observer was used for estimation purposes in order to improve the stability of the controller. The design and structure of the proposed pressure reducing valve affect the control performance was not considered in Li et al. (2017b). In underwater load applications, chattering at high frequency was reduced by using the adaptive backstepping sliding mode control (Liu et al., 2018). Sun et al. (2018) presented the ESO-based nonlinear controller in order to achieve a good transient response and position tracking in presence of parametric uncertainties. The classical PID controller was compared with the modified backstepping control, and it was determined that nonlinear control contributed better efficiency (Tri et al., 2015). A neural network-based estimator was a suitable technique used to improve the accuracy of the controller (Yao et al., 2018). Moreover, another radial basis function neural network-based nonlinear control was discussed (Zhu et al., 2018). Li et al. (2017a) used a fuzzy-based controller with a Cauchy sequene mutation as well as a mass weighing optimisation technique for a hydraulic turbine-based system. A fuzzy-based sliding mode controller with PSO was proposed and compared with the PID controller (Liang et al., 2017a). Table 14.2 summarises the details of nonlinear control methods for EHAS and further subsections present a detailed discussion.

### 14.3.2.1  Adaptive Control

In consideration of nonlinear dynamics, the adaptive control technique was deemed a suitable technique in more effective management in the presence of parametric uncertainty. An adaptive neural network-based control was designed for vibrating string systems and found globally stable (Zhao et al., 2018). Backlash type nonlinearity was reduced by using adaptive control in aircraft control, and stability was subsequently tested (Han et al., 2011). Yang et al. (2011) considered an industrial robot with uncertainties (i.e. friction parameter) and developed better position tracking by implementing an adaptive control structure based on backstepping. A novel adaptive control based on a neural network was proposed for continuous hot pressing EHS in the presence of input saturation and external disturbances in order to achieve tracking performance in the transient, as well as in steady-state, behaviour. However, measurement noise was not considered in considering stability. To

**TABLE 14.2**

**Research Category of Nonlinear Control**

| Research Category | Literature | Remarks |
|---|---|---|
| adaptive control | Han et al. (2011); Yang et al. (2011);Sun et al. (2012, 2015); Wang et al. (2015a); Yao et al. (2015b); Liu et al. (2016); Pan et al. (2017); Shen et al. (2017); Kang et al. (2018); Li and Wang (2018); Lyu et al. (2018); Zhao et al. (2018) | • Online parameters are updated using the adaptive law. <br> • Uncertainties and disturbances can suppress. <br> • Neural network, fuzzy and backstepping schemes can be applied with adaptive control. <br> • Improved stability and output accuracy ensue. |
| ESO-based control | Li et al. (2011); Yao et al. (2014); Guo et al. (2016); Wang et al. (2017); Shi et al. (2018); Li et al. (2019); Xu et al. (2019) | • Unmodeled disturbances and parametric uncertainties can estimate. <br> • Adaptive and backstepping controller-based ESO control improves output accuracy and the stability of the system. |
| robust control | Montaseri and Yazdanpanah (2012); Deng et al. (2015); Wang et al. (2015b); Gu et al. (2018); Cheng et al. (2019); Jing et al. (2019) | • Unmatched uncertainties and disturbances can be suppressed. <br> • Robustness of the controller improves. <br> • Chattering in response can be reduced and stability consequently increased. |

bring about better force tracking using adaptive control, the Maxwell-slip friction model was considered to compensate for uncertainties and disturbance (Kang et al., 2018). Lumped uncertainties were handled by the adaptive control law, which improves torque tracking accuracy, thus verifying stability (Li and Wang, 2018). Liu et al. (2016) described an adaptive fuzzy-based backstepping controller with a quantizer used to reduce the chattering problem. However, missing measurements and mismatched disturbances were left for further discussion. Throttling energy loss was reduced by applying adaptive robust control with good accuracy and stability (Lyu et al., 2018). Pan et al. (2017) highlighted the improvement in vertical dynamic performance in suspension systems by using an adaptive tracking controller. Further feasibility of adaptive control was discussed by Shen et al. (2017), Sun et al. (2012, 2015), Wang et al. (2015a) and Yao et al. (2015b) at various test platforms with different kind of uncertainties and nonlinearities.

### 14.3.2.2  ESO-Based Control

Extended state observer was used for state estimation as well as for parameter variations and unmodeled dynamics (Yao et al., 2014). Furthermore, a nonlinear backstepping controller was implemented and exhibited good robustness with asymptotic performance.

In two degrees of freedom with a robotic arm, some unmeasured states were estimated using ESO, and stabilities were categorised in three ways (Guo et al., 2016). Further stability was improved by applying backstepping control. Li et al. (2019) proposed a synchronisation control method based on the adaptive backstepping approach, and disturbances were estimated by using four ESOs. Mechanical and hydraulic dynamics uncertainties, as well controller parameters, were estimated using an ESO-based adaptive control scheme and exhibited good tracking control of single rod actuator systems (Wang et al., 2017). However, high feedback gain added instability in the controlled system. Moreover, the implementation of ESO was discussed, and good output response accuracy was identified for EHAS (Xu et al. 2019; Shi et al., 2018; Li et al., 2011).

### 14.3.2.3   Robust Control

The robust control technique is used to enhance the stability and robustness of the controller in the presence of nonlinearity and uncertainties. Cheng et al. (2019) discussed the proposed robust adaptive sliding mode controller, which was discovered more effective than the PID controller and adaptive backstepping controller for force loading the test platform. In addition, chattering in the force tracking response was reduced by applying the proposed robust controller. For a single input/single output system, a robust adaptive backstepping controller was implemented and produced asymptotic output tracking (Deng et al., 2015). In order to compensate for the dead-zone inverse model, a robust adaptive control was used, and further stability was improved by using the backstepping algorithm (Gu et al., 2018). Jing et al. (2019) applied a novel robust controller based on the radial basis function neural network to a continuous rotary system in the presence of external disturbances and parametric perturbation. In addition, an acceptable level of accuracy was determined in simulated results within the frequency band. The Lyapunov redesign method was used in rejecting unmatched uncertainty, and the robust adaptive controller improved the controller stability (Montaseri and Yazdanpanah, 2012). In Wang et al. (2015b), dual-loop control, which is comprised of velocity feed-forward compensator in open loop mode and closed loop robust torque controller was applied to the load simulator in the presence of nonlinearity, friction problems and disturbances due to actuator motion. In addition, the proposed controller was simple in structure and easily implemented.

## 14.4   CONCLUSION

This chapter conducted a general review about modelling and controlling of an electrohydraulic actuation system. Internal and external disturbances, parametric uncertainties and valve nonlinearities have been discussed. In addition, some assumptions were made in order to minimise the complexity in modelling of EHAS. According to the type of nonlinearity and uncertainty, various type of control schemes were deliberated. This study concludes that backstepping, adaptive control and ESO-based controllers, robust controllers and their derivatives were identified as the most suitable algorithms in controlling the EHAS system under nonlinearity. In addition, there is scope for future improvement in the control algorithm for time-varying internal and external disturbances, which can be measurable or non-measurable. Furthermore, accuracy and precision in tracking performance can be enhanced by developing new or hybrid controllers.

## REFERENCES

Ahn, Kyoung Kwan, Doan Ngoc Chi Nam, and Maolin Jin. 2013. "Adaptive backstepping control of an electrohydraulic actuator." *IEEE/ASME Transactions on Mechatronics* 19 (3): 987–995.

Asl, Reza Mohammadi, Rainer Palm, Huapeng Wu, and Heikki Handroos. 2020. "Fuzzy-based parameter optimization of adaptive unscented Kalman filter: Methodology and experimental validation." *IEEE Access* 8: 54887–54904.

Busquets, Enrique, and Monika Ivantysynova. 2015. "Discontinuous projection-based adaptive robust control for displacement-controlled actuators." *Journal of Dynamic Systems, Measurement, and Control* 137 (8).

Cerman, Otto, and Petr Hušek. 2012. "Adaptive fuzzy sliding mode control for electro-hydraulic servo mechanism." *Expert Systems with Applications* 39 (11): 10269–10277.

Chen, Shan, Zheng Chen, and Bin Yao. 2018. "Precision cascade force control of multi-DOF hydraulic leg exoskeleton." *IEEE Access* 6: 8574–8583.

Chen, Zhihuan, Yanbin Yuan, Xiaohui Yuan, Yuehua Huang, Xianshan Li, and Wenwu Li. 2015. "Application of multi-objective controller to optimal tuning of PID gains for a hydraulic turbine regulating system using adaptive grid particle swam optimization." *ISA Transactions* 56: 173–187.

Cheng, Lei, Zhen-Cai Zhu, Gang Shen, Shujing Wang, Xiang Li, and Yu Tang. 2019. "Real-time force tracking control of an electro-hydraulic system using a novel robust adaptive sliding mode controller." *IEEE Access* 8: 13315–13328.

Dang, Xuan Ba, DinhQuang Truong, Joonbum Bae, and Kyoung Kwan Ahn. 2019. "An effective disturbance-observer-based nonlinear controller for a pump-controlled hydraulic system." *IEEE/ASME Transactions on Mechatronics* 25 (1): 32–43.

Dell'Amico, Alessandro, and Petter Krus. 2015. "Modeling, simulation, and experimental investigation of an electrohydraulic closed-center power steering system." *IEEE/ASME Transactions on Mechatronics* 20 (5): 2452–2462.

Deng, Wenxiang, Jianyong Yao, and Dawei Ma. 2015. "Robust adaptive asymptotic tracking control of a class of nonlinear systems with unknown input dead-zone." *Journal of the Franklin Institute* 352 (12): 5686–5707.

Di Massa, Giandomenico, Riccardo Russo, Salvatore Strano, and Mario Terzo. 2013. "System structure identification and adaptive control of a seismic isolator test rig." *Mechanical Systems and Signal Processing* 40 (2): 736–753.

Dong, Yu, Jing Na, Shubo Wang, Guanbin Gao, and Haoran He. 2019. "Robust Adaptive Parameter Estimation and Tracking Control for Hydraulic Servo Systems." *In 2019 Chinese Control Conference (CCC)*, pp. 2468–2473. IEEE.

Ekoru, John E. D., and Jimoh O. Pedro. 2013. "Proportional-integral-derivative control of nonlinear half-car electro-hydraulic suspension systems." *Journal of Zhejiang University SCIENCE A* 14 (6): 401–416.

Gu, Weiwei, Jianyong Yao, Zhikai Yao, and Jingzhong Zheng. 2018. "Robust adaptive control of hydraulic system with input saturation and valve dead-zone." *IEEE Access* 6: 53521–53532.

Guo, Qing, Yi Zhang, Branko G. Celler, and Steven W. Su. 2016. "Backstepping control of electro-hydraulic system based on extended-state-observer with plant dynamics largely unknown." *IEEE Transactions on Industrial Electronics* 63 (11): 6909–6920.

Han, Kwang-Ho, Gi-Ok Koh, Jae-Min Sung, and Byoung-Soo Kim. 2011. "Adaptive control approach for improving control systems with unknown backlash." *In 2011 11th International Conference on Control, Automation and Systems*, pp. 1919–1923. IEEE.

Ijaz, Salman, Lin Yan, Mirza Tariq Hamayun, Waqas Mehmood Baig, and Cun Shi. 2018. "An adaptive LPV integral sliding mode FTC of dissimilar redundant actuation system for civil aircraft." *IEEE Access* 6: 65960–65973.

Jing, Wang X., Liu M. Zhen, Sun Y. Wei, and Xin Wang. 2019. "Research on low-speed performance of continuous rotary electro-hydraulic servo motor based on robust control with Adaboost prediction." *The Journal of Engineering* 2019 (13): 60–67.

Kang, Shuo, Hao Yan, Lijing Dong, and Changchun Li. 2018. "Finite-time adaptive sliding mode force control for electro-hydraulic load simulator based on improved GMS friction model." *Mechanical Systems and Signal Processing* 102: 117–138.

Kayacan, Erdal, and Okyay Kaynak. 2012. "Sliding mode control theory-based algorithm for online learning in type-2 fuzzy neural networks: application to velocity control of an electro hydraulic servo system." *International Journal of Adaptive Control and Signal Processing* 26 (7): 645–659.

Kumar, Rajesh, Smriti Srivastava, and J. R. P. Gupta. 2017. "Diagonal recurrent neural network based adaptive control of nonlinear dynamical systems using lyapunov stability criterion." *ISA Transactions* 67: 407–427.

Li, Yunhua, and Liuyu He. 2015. "Counterbalancing speed control for hydrostatic drive heavy vehicle under long down-slope." *IEEE/ASME Transactions on Mechatronics* 20 (4): 1533–1542.

Li, Chaoshun, Yifeng Mao, Jianzhong Zhou, Nan Zhang, and Xueli An. 2017a. "Design of a fuzzy-PID controller for a nonlinear hydraulic turbine governing system by using a novel gravitational search algorithm based on Cauchy mutation and mass weighting." *Applied Soft Computing* 52: 290–305.

Li, Shihua, Jun Yang, Wen-Hua Chen, and Xisong Chen. 2011. "Generalized extended state observer based control for systems with mismatched uncertainties." *IEEE Transactions on Industrial Electronics* 59 (12): 4792–4802.

Li, Ting, Ting Yang, Yuyan Cao, Rong Xie, and Xinmin Wang. 2019. "Disturbance-estimation based adaptive backstepping fault-tolerant synchronization control for a dual redundant hydraulic actuation system with internal leakage faults." *IEEE Access* 7: 73106–73119.

Li, Yong, and Qingfeng Wang. 2018. "Pump-pressure-compensation-based adaptive neural torque control of a hydraulic excavator with open center valves." *In 2018 IEEE/ASME International Conference on Advanced Intelligent Mechatronics (AIM)*, pp. 610–615. IEEE.

Li, Yangyang, Peng Zhang, Dong Li, Yunhua Li, and Liman Yang. 2017b. "Backstepping adaptive control of dual-variable electro-hydraulic actuator with displacement-pressure regulation pump." *In 2017 12th IEEE Conference on Industrial Electronics and Applications (ICIEA)*, pp. 1206–1211. IEEE.

Liang, Ji, Xiaohui Yuan, Yanbin Yuan, Zhihuan Chen, and Yuanzheng Li. 2017. "Nonlinear dynamic analysis and robust controller design for Francis hydraulic turbine regulating system with a straight-tube surge tank." *Mechanical Systems and Signal Processing* 85: 927–946.

Liu, Heng, Haojie Li, Qingfang Hou, Chenguang Xu, and Jianxing Leng. 2018. "A pressure-displacement control for an active-passive hybrid compensation system." *In 2018 OCEANS-MTS/IEEE Kobe Techno-Oceans (OTO)*, pp. 1–6. IEEE.

Liu, Wenhui, Cheng-Chew Lim, Peng Shi, and Shengyuan Xu. 2016. "Backstepping fuzzy adaptive control for a class of quantized nonlinear systems." *IEEE Transactions on Fuzzy Systems* 25 (5): 1090–1101.

Lu, Peng, Timothy Sandy, and Jonas Buchli. 2019. "Adaptive unscented Kalman filter-based disturbance rejection with application to high precision hydraulic robotic control." *In*

*2019 IEEE/RSJ International Conference on Intelligent Robots and Systems (IROS)*, pp. 4365–4372. IEEE.

Lyu, Litong, Zheng Chen, and Bin Yao. 2018. "High precision and high efficiency control of pump and valves combined hydraulic system." *In 2018 IEEE 15th International Workshop on Advanced Motion Control (AMC)*, pp. 391–396. IEEE.

Mintsa, HonorineAngue, Ravinder Venugopal, Jean-Pierre Kenne, and Christian Belleau. 2011. "Feedback linearization-based position control of an electrohydraulic servo system with supply pressure uncertainty." *IEEE Transactions on Control Systems Technology* 20 (4): 1092–1099.

Montaseri, Ghazal, and Mohammad Javad Yazdanpanah. 2012. "Adaptive control of uncertain nonlinear systems using mixed backstepping and Lyapunov redesign techniques." *Communications in Nonlinear Science and Numerical Simulation* 17 (8): 3367–3380.

Pan, Huihui, Weichao Sun, Xingjian Jing, Huijun Gao, and Jianyong Yao. 2017. "Adaptive tracking control for active suspension systems with non-ideal actuators." *Journal of Sound and Vibration* 399: 2–20.

Pasolli, Philipp, and Michael Ruderman. 2018. "Linearized piecewise affine in control and states hydraulic system: modeling and identification." *In IECON 2018-44th Annual Conference of the IEEE Industrial Electronics Society*, pp. 4537–4544. IEEE.

Shen, Gang, Xiang Li, Zhencai Zhu, Yu Tang, Weidong Zhu, and Shanzeng Liu. 2017. "Acceleration tracking control combining adaptive control and off-line compensators for six-degree-of-freedom electro-hydraulic shaking tables." *ISA Transactions* 70: 322–337.

Shen, Gang, Zhen-Cai Zhu, Xiang Li, Yu Tang, Dong-Dong Hou, and Wen-Xiang Teng. 2016. "Real-time electro-hydraulic hybrid system for structural testing subjected to vibration and force loading." *Mechatronics* 33: 49–70.

Shi, Wenzhuo, Jianhua Wei, and Jinhui Fang. 2018. "Desired compensation nonlinear cascade control of high-response proportional solenoid valve based on reduced-order extended state observer." *IEEE Access* 6: 64503–64514.

Sotiropoulos, Filippos E., and H. Harry Asada. 2019. "A model-free extremum-seeking approach to autonomous excavator control based on output power maximization." *IEEE Robotics and Automation Letters* 4 (2): 1005–1012.

Strano, Salvatore, and Mario Terzo. 2016. "Accurate state estimation for a hydraulic actuator via a SDRE nonlinear filter." *Mechanical Systems and Signal Processing* 75: 576–588.

Sui, Shuai, and Shaocheng Tong. 2016. "Fuzzy adaptive quantized output feedback tracking control for switched nonlinear systems with input quantization." *Fuzzy Sets and Systems* 290: 56–78.

Sun, Chungeng, Jinhui Fang, Jianhua Wei, and Bo Hu. 2018. "Nonlinear motion control of a hydraulic press based on an extended disturbance observer." *IEEE Access* 6: 18502–18510.

Sun, Weichao, Huihui Pan, and Huijun Gao. 2015. "Filter-based adaptive vibration control for active vehicle suspensions with electrohydraulic actuators." *IEEE Transactions on Vehicular Technology* 65 (6): 4619–4626.

Sun, Weichao, Zhengli Zhao, and Huijun Gao. 2012. "Saturated adaptive robust control for active suspension systems." *IEEE Transactions on industrial electronics* 60 (9): 3889–3896.

Todeschini, Fabio, Matteo Corno, Giulio Panzani, Simone Fiorenti, and Sergio M. Savaresi. 2014. "Adaptive cascade control of a brake-by-wire actuator for sport motorcycles." *IEEE/ASME Transactions on Mechatronics* 20 (3): 1310–1319.

Tri, Nguyen Minh, Doan Ngoc Chi Nam, HyungGyu Park, and Kyoung Kwan Ahn. 2015. "Trajectory control of an electro hydraulic actuator using an iterative backstepping control scheme." *Mechatronics* 29: 96–102.

Vaezi, Masoud, and Afshin Izadian. 2015. "Piecewise affine system identification of a hydraulic wind power transfer system." *IEEE Transactions on Control Systems Technology* 23 (6): 2077–2086.

Wang, Xiao, Dacheng Cong, Zhidong Yang, Shengjie Xu, and Junwei Han. 2019. "Modified Quasi-Newton optimization algorithm-based iterative learning control for multi-axial road durability test rig." *IEEE Access* 7: 31286–31296.

Wang, Chengwen, Zongxia Jiao, and Long Quan. 2015a. "Adaptive velocity synchronization compound control of electro-hydraulic load simulator." *Aerospace Science and Technology* 42: 309–321.

Wang, Chengwen, Zongxia Jiao, and Long Quan. 2015b. "Nonlinear robust dual-loop control for electro-hydraulic load simulator." *ISA Transactions* 59: 280–289.

Wang, Chengwen, Zongxia Jiao, Shuai Wu, and Yaoxing Shang. 2014a. "A practical nonlinear robust control approach of electro-hydraulic load simulator." *Chinese Journal of Aeronautics* 27 (3): 735–744.

Wang, Chengwen, Zongxia Jiao, Shuai Wu, and Yaoxing Shang. 2014b. "Nonlinear adaptive torque control of electro-hydraulic load system with external active motion disturbance." *Mechatronics* 24 (1): 32–40.

Wang, Chengwen, Long Quan, Zongxia Jiao, and Shijie Zhang. 2017. "Nonlinear adaptive control of hydraulic system with observing and compensating mis-matching uncertainties." *IEEE Transactions on Control Systems Technology* 26 (3): 927–938.

Won, Daehee, Wonhee Kim, Donghoon Shin, and Chung Choo Chung. 2014. "High-gain disturbance observer-based backstepping control with output tracking error constraint for electro-hydraulic systems." *IEEE Transactions on Control Systems Technology* 23 (2): 787–795.

Xia, Yingkai, Kan Xu, Ye Li, Guohua Xu, and Xianbo Xiang. 2018. "Modeling and three-layer adaptive diving control of a cable-driven underwater parallel platform." *IEEE Access* 6: 24016–24034.

Xu, Zhangbao, Lan Li, Jianyong Yao, Xiaolei Hu, Qingyun Liu, and Nenggang Xie. 2019. "State constraint control for uncertain nonlinear systems with disturbance compensation." *IEEE Access* 7: 155251–155261.

Xudong, Li, Chen Xiong, Yao Jianyong, Xu Jinsheng, and Zhou Changsheng. 2018. "Nonlinear robust motion control of hydraulic servo system with matched and mismatched disturbance compensation." *In 2018 Chinese Control And Decision Conference (CCDC)*, pp. 1686–1691. IEEE.

Yao, Jianyong, Wenxiang Deng, and Zongxia Jiao. 2015a. "Adaptive control of hydraulic actuators with LuGre model-based friction compensation." *IEEE Transactions on Industrial Electronics* 62 (10): 6469–6477.

Yao, Jianyong, Zongxia Jiao, and Dawei Ma. 2015b. "A practical nonlinear adaptive control of hydraulic servomechanisms with periodic-like disturbances." *IEEE/ASME Transactions on Mechatronics* 20 (6): 2752–2760.

Yang, Chifu, Shutao Zheng, Xinjie Lan, and Junwei Han. 2011. "Adaptive robust control for spatial hydraulic parallel industrial robot." *Procedia Engineering* 15: 331–335.

Yao, Jianyong, Zongxia Jiao, and Dawei Ma. 2014. "Extended-state-observer-based output feedback nonlinear robust control of hydraulic systems with backstepping." *IEEE Transactions on Industrial Electronics* 61 (11): 6285–6293.

Yao, Zhikai, Jianyong Yao, and Weichao Sun. 2018. "Adaptive RISE control of hydraulic systems with multilayer neural-networks." *IEEE Transactions on Industrial Electronics* 66 (11): 8638–8647.

Ye, Ning, Jinchun Song, and Guangan Ren. 2020. "Model-based adaptive command filtering control of an electrohydraulic actuator with input saturation and friction." *IEEE Access* 8: 48252–48263.

Ye, Yi, Chen-Bo Yin, Yue Gong, and Jun-jing Zhou. 2017. "Position control of nonlinear hydraulic system using an improved PSO based PID controller." *Mechanical Systems and Signal Processing* 83: 241–259.

Yin, Xiuxing, and Xiaowei Zhao. 2019. "Sensorless maximum power extraction control of a hydrostatic tidal turbine based on adaptive extreme learning machine." *IEEE Transactions on Sustainable Energy* 11 (1): 426–435.

Zhang, Dan, and Bin Wei. 2017. "A review on model reference adaptive control of robotic manipulators." *Annual Reviews in Control* 43: 188–198.

Zhao, Jian, Dongjian Song, Bing Zhu, Zhicheng Chen, and Yuhang Sun. 2020. "Nonlinear backstepping control of electro-hydraulic brake system based on bond graph model." *IEEE Access* 8: 19100–19112.

Zhao, Zhijia, Jun Shi, Xuejing Lan, Xiaowei Wang, and Jingfeng Yang. 2018. "Adaptive neural network control of a flexible string system with non-symmetric dead-zone and output constraint." *Neurocomputing* 283: 1–8.

Zhao, Chun, Cungui Yu, and Jianyong Yao. 2019. "Dynamic decoupling based robust synchronous control for a hydraulic parallel manipulator." *IEEE Access* 7: 30548–30562.

Zhu, Liangkuan, Zibo Wang, Yugang Zhou, and Yaqiu Liu. 2017. "Adaptive neural network saturated control for MDF continuous hotpressing hydraulic system with uncertainties." *IEEE Access* 6: 2266–2273.

Zhu, Liangkuan, Yugang Zhou, and Yaqiu Liu. 2018. "Robust adaptive neural prescribed performance control for MDF continuous hot pressing system with input saturation." *IEEE Access* 6: 9099–9113.

# 15 Study of Mechanical Properties of Photo Polymer Lab-on-a-Chip Structures Fabricated by Scan-Based Microstereolithography

*K. Raghavendra, M Manjaiah, K Ankit, and N Balashanmugam*

## CONTENTS

## 15.1   INTRODUCTION

Rapid prototyping (RP) is one of the far-flung terms used to describe the technology of additive manufacturing systems. For decades, RP has played a vital role in the development of functional structures that are complex in design and difficult to conventionally manufacture (Zhang et al. 1999). The first commercialised machine tool based on additive manufacturing was built in 1980. It was essentially used to do rapid prototyping work. With the advent of rapid prototyping, the manufacturing of micro products with typical shapes and features became possible. Equipped with the knowledge of the advantages and disadvantages of various 3D printing techniques, models can be developed for predefined properties depending on the material implemented for the development of the device for any specific application and functionality. Any purpose of a complex structure can be well defined, if the system is made simple to use. For this reason, researchers (Zhang et al. 1999) developed the most straightforward process in additive manufacturing technology, stereolithography. Stereolithography is a technology that combines various disciplines of science and engineering to form a new potential in processing prototypes (Bertsch et al. 2000). Through this process, it was surmounted that it is possible to mould a blend of polymers to structures of any size and shape, which can be then applied to various industrial sectors such as nuclear, biomedical, and automotive industries, etc (Bertsch et al. 2001).

In the present work, the principle investigation focuses on studying the effects of various process parameters of scan-based microstereolithography (SMSL) on the overall mechanical characteristics of the structure in development. Attempts to understand the mechanical behaviour of the structures developed by scan-based microstereolithography systems using ASTM E2546-07, also known as nanoindentation, have been thusly made (Hay et al. 1999). Surface and dimensional integrity studies have been carried out by using an optical profiler and scanning electron microscope. The aim of the authors is to study the change in values of Young's modulus and hardness with a change in process parameters.

## 15.2   LAYER-BY-LAYER FABRICATION OF TEST SPECIMENS FOR CHARACTERISATION

Some techniques are specifically meant to focus on the experimental relevance and complexity of the system functionality, while others ideally concentrate on manufacturability. However, specific applications demand a balance to preserve manufacturability with no compromise in quality in order to manufacture a cost-efficient, automated and user-friendly system which can be part of research labs and used for social welfare. In addition, though the ability to manufacture exact replicates of the design is extremely attractive and challenging, most of the technology to achieve that stage in the micro-level requires high precision with low scope for errors.

Further attempts are still required to complement functionalities for the replacement of high-cost scientific modules requiring upgraded technology in the domain of micro and NANO technology for studies related to testing system modelling relevant to lab-on-a-chip (LOC) development. Since micro-level test procedures are exceedingly complex, the degree to which researchers must control the analysis microenvironment in order to elicit desired test response is still largely unclear. Hence, there will be a prevailing process that involves model construction and experimental validation to standardise the process parameters in the micro testing regime for future application in LOC.

The current study focuses on using microstereolithographic 3D printing techniques to manufacture components deployable in LOC, while mechanically stable photopolymer is used to construct a 3D self-mixing vortex flow for the microfluidic channel, where a micro impellor is designed to act as a super mixer, as illustrated in Figure 15.1 a-b. The goal of the study is further subdivided into the following objectives:

1. develop a suitable photopolymer resin for printing LOC components for mechanically stable photopolymer structures of relevant sizes
2. establish optimum process parameters for steady and stable LOC components with various mechanical and chemical properties requirements

## 15.2.1 LAB-ON-A-CHIP (LOC)

A lab-on-a-chip is a miniaturised device that integrates onto a single chip one or several analyses which are usually done in a laboratory setting that generally focuses on research relevant to the synthesis of chemicals (Mirasoli et al. 2014). It is used due to various advantages, such as cost efficiency, parallelisation and ergonomic suitability. Diagnostic specifically relies on two core technologies: Microfluidics and molecular biology.

One of the main components in LOC is the microfluidic channel chip has a set of interconnected microchannels that are etched or moulded into a type of material (e.g. glass, silicon or polymer). This feature can be used in various electronic applications as a cooling device of electronic components (Gong et al. 2017).

## 15.3 EXPERIMENTAL SETUP

In studies done to develop LOC components, a scan-based microstereolithography setup was indigenously instituted. The schematic and photograph of the setup are as shown in Figure 15.1. An Argon Ion UV laser with a wavelength of 364 nm was used as the light source, and a UV coherent lens with a depth of focus (DOF) or focal length of 50.3 mm used to focus the beam onto the resin surface was adopted. The other parts of the system comprised of: an acoustic optic modulator (AOM) used to achieve a highly efficient diffracted (first-order) beam for effective utilisation of UV laser power.

A set of optical mirrors and an optical iris diaphragm (1.5 mm – 11 mm aperture range, ten leaves, Newport USA) were made to guide the beam towards the resin surface with the highest possible power and lowest spot size of the beam. An X-Y-Z linear translation stage of 20 nm metre resolution is connected to a motion controller drive (Model-XPS, Newport). This linear stage moves the substrate in three orthogonal directions within the photopolymer container irradiated to a fixed UV beam for achieving the programmed structure.

(a)

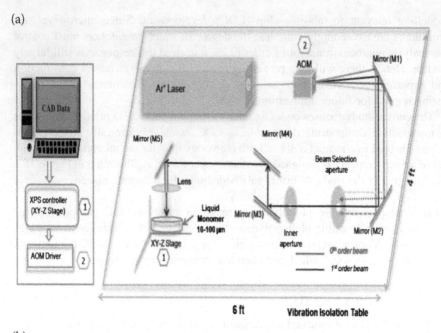

(b)

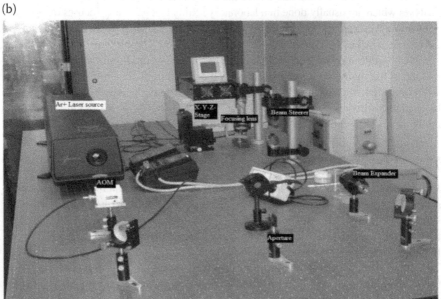

**FIGURE 15.1** Experimental Setup for Scan-Based Microstereolithography (a) Schematic Representation (b) Actual Setup.

## 15.3.1 PREPARATION OF PHOTOPOLYMER FOR UV CURING

As procured in precursor preparation, 1,6-hexanediol diacrylate (1,6-HDDA) as the monomer and benzoin ethyl ether (BEE) as the photoinitiator, obtained from Sigma Aldrich (USA), was used. The monomer PI blend of different concentrations was applied to investigate the effect of UV laser on the final structure for fixed process parameters. In the typical procedure followed for the preparation of the photocurable solution, 1,6-HDDA was combined with BEE with a fixed volumetric relation. The PI monomer solution was then stirred using a magnetic stirrer at room temperature in a dark room environment for a duration of four hours.

## 15.4  FABRICATIONS AND TESTING

### 15.4.1 DEVELOPMENT OF PHOTOPOLYMER SOLUTION

The design and initial code generation of structures were made using commercial CAD software (i.e. UG NX5). The codes were then post-processed and made compatible with LABVIEW 8.5.1 software. The controlling stage and the AOM are interfaced with LABVIEW 8.5.1 for path generation using preprocessed codes. Laser fabrication parameters such as power and laser spot diameter were optimised on initial solidification trials (Manias et al. 2001; Prasanna et al. 2013). While maintaining the laser beam on a fixed spot, the silicon substrate placed on the substrate holder fastened to the linear translational stage is dipped into the vat containing liquid polymer and is subjected to predefine scanning in order to obtain the designed structure. The crosslink scheme of the polymer structure when exposed to a UV beam of 364 nm wavelength is shown in Figure 15.2.

### 15.4.2 SELECTION OF PROCESS PARAMETERS

#### 15.4.2.1 Laser Exposure

For the fabrication of the test specimen and other application-based structures, an ultraviolet (UV) laser beam of 20 μm spot diameter was used. Since the process involves multiple line polymerisation in the fabrication of the desired cross-section area, single line and point solidification studies are conducted using the microscopic images of the line scanned by microstereolithography for different laser energy exposures (Figure 15.3). Theoretical comparisons using Jacob's equations of the tested lines and points were then conducted for numerical closeness.

#### 15.4.2.2 Scan Pattern and Layer Thickness

In the fabrication of stable structures using SMSL, it is essential to study the effect of scan pattern and layer thickness during solidification. Figure 15.4 shows the fabrication of simple geometry using different scan patterns. The 3D CAD model is illustrated in Figure 15.4 with the following information: For circular geometry, the diameter, scan pitch and layer thickness were 2,000 μm, 30 μm and 50 μm respectively. For square geometry, the cross-section area, scan pitch and layer

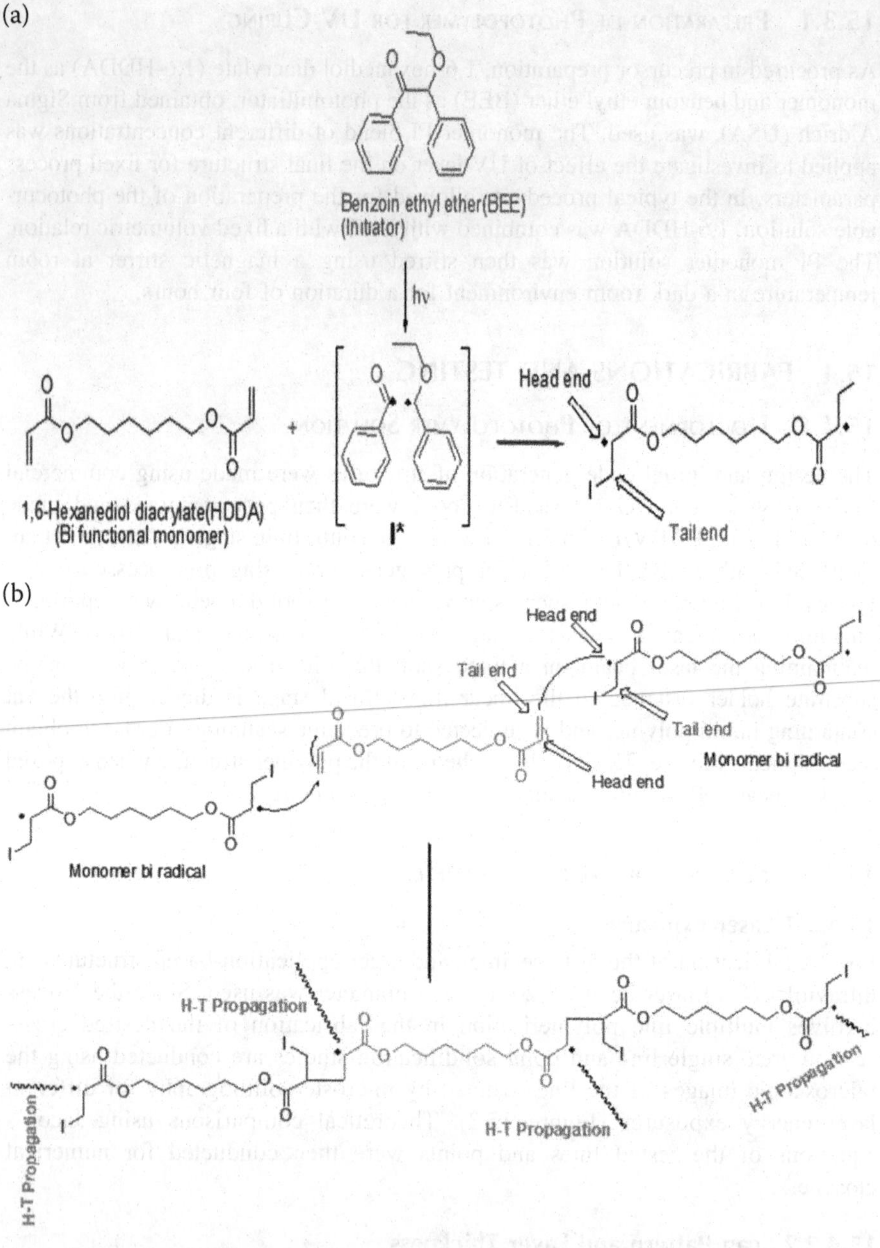

(a)

Benzoin ethyl ether(BEE)
(Initiator)

1,6-Hexanediol diacrylate(HDDA)
(Bifunctional monomer)

Head end

Tail end

(b)

Head end

Tail end

Tail end

Monomer bi radical

Head end

Monomer bi radical

H-T Propagation

H-T Propagation

H-T Propagation

H-T Propagation

**FIGURE 15.2** Crosslink Scheme of the Polymer Blend (a) Initiation (b) Propagation (c) Termination.

thickness were 4,000 μm², 30 μm and 50 μm respectively. A layer thickness of 20 μm, 75 μm and 100 μm were also used in this study to assess the effect of layer thickness on the material properties of the structure.

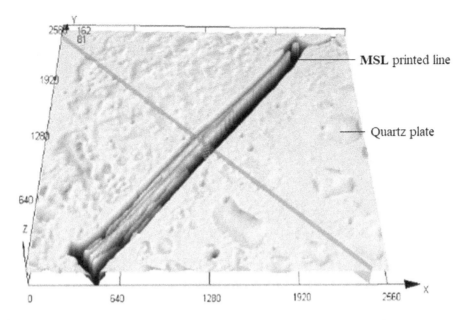

**FIGURE 15.3**  Microscopic Image of the Line Scanned Using Microstereolithography for a Single Scan Exposure Test over Quartz Plate Using Confocal Microscope.

## 15.4.2.3 Scan Orientation

As studied in conventional laminate design practice, the directional nature of stacking layers provides a material the ability to withstand high loads without failure (Yu et al. 2006). In this context, a study has been made by fabricating structures with multiple stacking orientations of layers to enhance mechanical properties. As shown in Figures 15.5 and 15.6, the stacking of layers with standard scan orientations of 0/±45/90 was laid in a different order with four to six layers of 20 μm thickness for each specimen. Investigations were carried out to make a comprehensive overview of the structural testing of polymer materials using the nanoindentation technique. This study will provide a thorough knowledge of the effect of layer stacking with different orientations on the mechanical characteristics of the polymer structure.

## 15.4.3  NANOINDENTATION TEST PROTOCOLS

The experiments of nanoindentation were carried out using the Agilent nanoindenter G-200 (Agilent Tech, USA) which has a load resolution of 1 nm. In the current study, a constant stiffness measurement technique was used. In this technique, the instantaneous elastic behaviour of a material is documented based on the area of contact of the indenter. Furthermore, for the area of contact of the indenter, the modulus and hardness are determined as functions of indentation penetration depth following a single load/unload cycle.

(a)

(b)

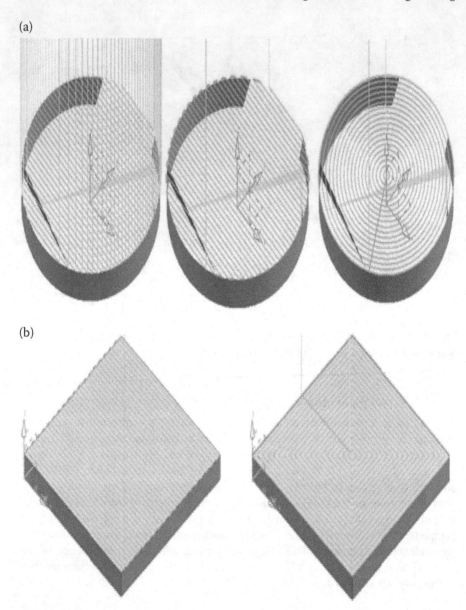

**FIGURE 15.4** CAD Model of Simple Geometry for Different Scan Patterns (a) Circular Geometry with Zig-Zag, Follow Path and Zig with Contour Scan Pattern (b) Square Geometry with Zig-Zag and Follow Part Scan Pattern.

The indentation procedure was carried out using a Diamond point Berkowich indenter in the following order:

    i. The surface approach velocity was set to 5 nm/s at a tip specimen gap of 1,000 nm.

(a)                          (b)                          (c)

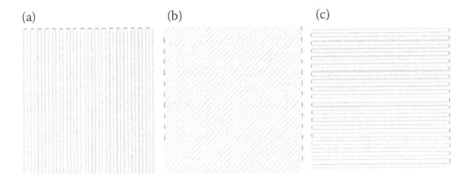

**FIGURE 15.5**   CAD Design of Layers for Different Orientations (a) 90° (b) 45° (c) 0°.


ii. A constant strain rate of 0.05 s$^{-1}$ on the target was maintained throughout the maximum depth of penetration at 5,000 nm.

iii. Five indents were made on each sample with unchanged loading conditions. The depth of indentation was maintained to 8% of the overall thickness of the sample in order to avoid the substrate effect.

## 15.5   RESULTS AND DISCUSSIONS

### 15.5.1   OPTICAL PROFILER FOR SURFACE ROUGHNESS MEASUREMENTS

Surface roughness measurements were carried out using optical profiler for samples with varied scan pitch and layer stack patterns. More than 30 different parameter variations were made. The measured data of surface roughness discovered as diverse with process parameters. When the scan pitch was increased from 0.75d pitch to 1.5d pitch distance, a moderate change in the surface roughness value was observed, with the optimised value of layer thickness at 20 µm.

Surface roughness for different classes of scan orientations (0°/±45°/90°) was measured. As can be identified, the negligible deflection of the values in surface roughness was documented as the scan pitch and exposure time for all of the orientations were kept constant.

(a)                          (b)                          (c)

**FIGURE 15.6**   Fabricated Structures for Different Scan Patterns Circular Geometry Fabricated with (a) Zig Zag, (b) Follow Part and (c) Zig with Contour Scan Pattern.

## 15.5.2 Effect of Laser Exposure

With optimised chemical parameters, the process was examined for the effect of laser power on the geometrical profile of the structure. In this context, curing depth ($C_d$) and curing width ($C_w$) studies were made for theoretical and experimental evaluation. From the Equations (15.1) and (15.2) used for calculating $C_d$ and $C_w$, it was evident that an increase in power consequently raised the value of $C_d$ and $C_w$ (Jacobs 1992).

$$C_d = D_P \ln\left[\frac{E_{max}}{E_C}\right]$$  (15.1)

$$C_w = B\sqrt{\frac{C_d}{2D_p}}.$$  (15.2)

The indentation platform was used to determine the hardness values of the samples. Standard indentation load was measured as a function of deformation depth (Figure 15.7a), and Berkovich type diamond indenter (Figure 15.7b) was used. A 30 mN maximum load in the experiment was performed in the following sequence: (1) Approaching the surface, (2) loading to the peak load of 30 mN at the rate of 5 mN/s, (3) holding time of the indenter with a peak load for 30 s, (4) unloading from the peak load at the rate of 5 mN/s. Finally, hardness value plots were obtained through the analysis by Agilent software.

With optimised chemical parameters, the experimental evaluation of laser power on the geometrical profile of the structure was studied. As shown in Figure 15.7(c), geometrical profile studies were made to evaluate the inconsistency of surface profile as well as mechanical properties for a scan speed of 1 mm/min over laser power of 8 µW and 9 µW. For every change in power, geometrical undulation was found on the sample surface, with variation in line width (Cw) and depth (Cd) as well as considerable changes in the mechanical characteristics of the samples, as shown in Figure 15.7(c). Hence, after characterising the mess structure for geometrical stability at power input 10 µW (Figure 15.8) power was considered to be an optimal value in the development of structures for different applications with stabilised load distribution.

When compared to low power (8 µW – 9 µW), the increased power intensity (10 µW) prints (Figure 15.8) show a better surface profile and improved mechanical properties. Hence at 10 µW, the surface was made of a complete layer that was photopolymerized by laser beams with no major geometrical undulations, thus leading to better geometrical and hardness properties. On the other hand, the surface in other power inputs consists mostly of semi-reacted layers leading to uncured samples.

## 15.5.3 Nanoindentation Test Results

Experiments were carried out according to the standard procedure as mentioned in ASTM E256. Indentation depth was maintained at a maximum constant depth of penetration of 5,000 nm in order to prevent the polymer from adhering to the

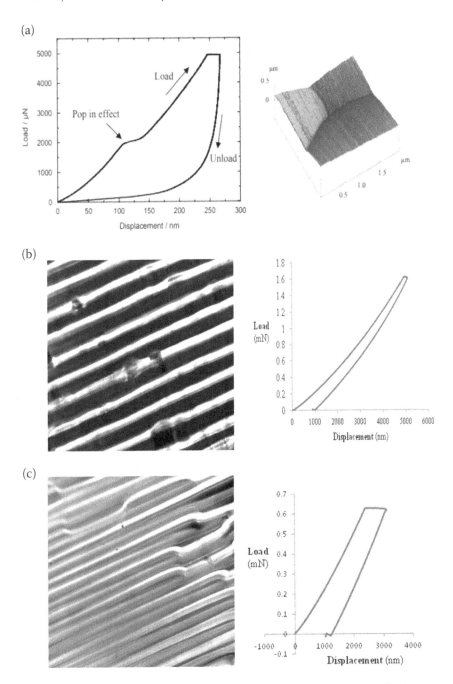

**FIGURE 15.7** Load against Displacement for Nanoindentation (a) Load versus Displacement Curve during Test (b) Residual Impression after Indentation (Neumeister et al. 2008) (Reprinted from *Journal of Laser Micro/Nanoengineering*, 3 (2), Neumeister, A., 67–72, 2008, with kind permission of JLPS-Japan Laser Processing Society) (c) Load distribution at 8 μW and 9 μW with surface profiles.

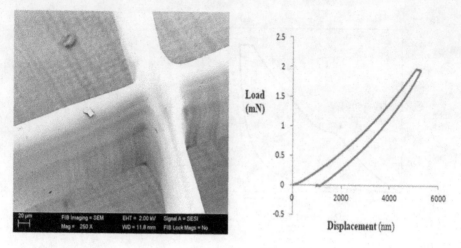

**FIGURE 15.8**   SEM Image of Mess Structure Fabricated at 10 μW Power with all Other Process Parameters Optimised.

surface of the indenter. The indentation process exhibited different results for different conditions. The indentation test carried out on the polymer samples shows no residual indentation marks. From this result, it is evident that the material behaves viscoelastically when processed using scan-based microstereolithography due to limitations in laser energy to which it is exposed.

The results of the modulus and hardness tested in the current study varied from specimen to specimen with diverse process parameters. For varied scan pitches of 0.75d, 1d and 1.25d, outcomes were discovered as non-repetitive at different points on the surface of the sample. However, with a scan pitch value of 1.5d, the values were quite repetitive with every sample fabricated. The reason for such specimen behaviour at lower pitch values was due to the overlapping of scanned lines which turns the layer elastically soft, leading to changes in mechanical properties at different points.

To study the effects of laser beam scanning patterns for proper solidification of photopolymer, fabrication samples were programmed for 0°/±45°/95° vector scan angles. As shown in Figure 15.5 and Figure 15.6, the structures were also tested for effective formation with different scan patterns. Results obtained from the SEM characterisation show the zigzag pattern as performing satisfactorily for all considered geometries of structures (Figure 15.10). With zig-zag as an optimised scan pattern, the samples were then fabricated for different scan angles 0°/±45°/90° with a layer thickness of 20 μm.

As shown in Table 15.1 and Figure 15.9, the following observations were made when characterising mechanical properties using nanoindentation with different layer stacking:

1. For layers with mono orientations, the material was discovered to be highly viscoelastic due to a unidirectional layup. Hence, hardness and modulus values were hard to determine.

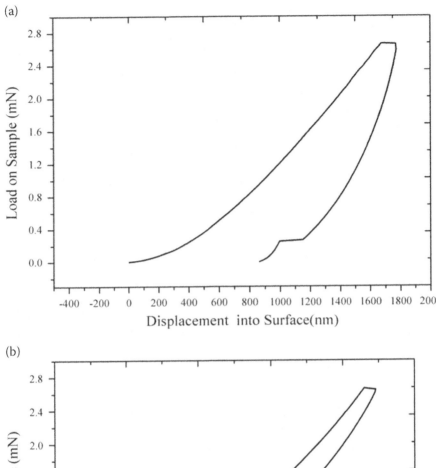

**FIGURE 15.9** Load versus displacement Curves for Different Scan Orientations of 0°+90°-45°+45°-90°-0° Orientations Sample Surface (a) Displacement Limited to 2,000 nm (b) Displacement limited to 5,000 nm.

(a)

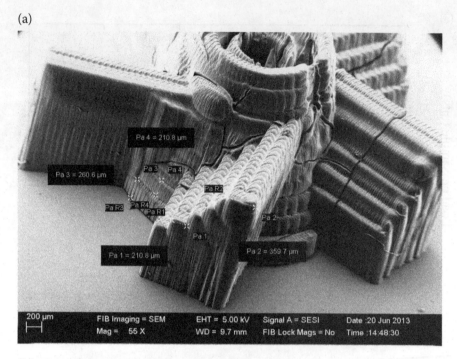

(b)

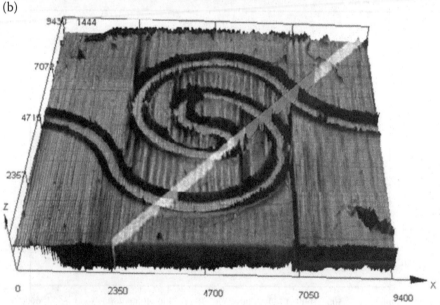

**FIGURE 15.10** Complex 3-Dimensional Structures Developed for Use in Microsystem Applications (a) 3D Micro Impeller (b) Vortex Flow Microfluidic Channel.

**TABLE 15.1**
**Modulus and Hardness of Samples with Different Scan Pitch AND Pattern Orientations**

| Sl. No | Pitch (%) | Degree of Orientations | Modulus (GPa) | Hardness (GPa) |
|--------|-----------|------------------------|---------------|----------------|
| 1 | 75 | 90 | 0.007 | 0.002 |
|   |    | 0, 90 | 0.012 | 0.003 |
|   |    | 45, 90 | 0.014 | 0.005 |
|   |    | -45, 45 | 0.017 | 0.005 |
| 2 | 100 | 90 | X | X |
|   |    | 0, 90 | X | X |
|   |    | 45, 90 | 0.017 | 0.004 |
|   |    | -45, 45 | 0.018 | 0.006 |
| 3 | 125 | 90 | 0.02 | 0.007 |
|   |    | 0, 90 | 0.019 | 0.006 |
|   |    | 45, 90 | X | X |
|   |    | -45, 45 | X | X |
| 4 | 150 | 90 | 0.022 | 0.008 |
|   |    | 0, 90 | 0.021 | 0.009 |
|   |    | 45, 90 | 0.023 | 0.011 |
|   |    | -45, 45 | 0.022 | 0.01 |

2. For stacking with two different orientations, an insignificant difference was found in the material properties. However, the values of hardness were found on certain areas of the specimen, indicating the effect of scan orientation on the material property. Also, the values of hardness were determined to be high at the corners due to the high dwell time of laser exposure.

3. For multiple combinations of scan orientations of $0°/\pm45°/90°$, the behaviour of the structure's mechanical characteristic was elastoplastic with repeatability in the values of the hardness of elastic modulus and hardness.

4. Maximum mechanical strength was attained with the scan orientation of $0°+90°-45°+45°-90°-0°$. The cause for increased mechanical property in the structure may be due to the addition of fibres in the transverse direction, thus strengthening the longitudinal direction fibres. This, in turn, increases the mechanical properties of the microstructure in production.

## 15.5.4 LOC COMPONENTS DEVELOPED AFTER PROCESS OPTIMISATION

The scan-based microstereolithography (SMSL) system has been successfully implemented for the fabrication of structures that are complex in design and development for various applications. The authors hitherto have presented a few examples of such structures that can be used in micro applications. Figure 15.10 shows the SEM image and confocal microscopic images of the 3D impeller and vortex flow microfluidic channel respectively developed with the SMSL system at optimised process parameters.

## 15.6 CONCLUSIONS

This chapter presented a comprehensive study conducted by the authors in defining the capability of the scan-based microstereolithography system in the development of structures that are mechanically, geometrically and dimensionally stable. An indigenously developed microstereolithography setup performed adequately, and a photopolymer solution was developed. Furthermore, optimum parameters of microstereolithography were obtained and used to make structures, and finally, to make complex 3D structures to be used in lab-on-a-chip and microfluidics applications. In conclusion, this study has identified and presented that scan-based microstereolithography is capable to fabricate devices for biomedical applications.

## REFERENCES

ASTM E2546-07. Standard Method for Instrumented Indentation Testing, 2014. ASTM International, West Conshohocken, PA, USA. www.astm.org.
Bertsch, A., P. Bernhard, and P. Renaud. 2001. Microstereolithography: concepts and applications. *8th International Conference on Emerging Technologies and Factory Automation. Proceedings* 2: 289–298.
Bertsch, A., P. Bernhard, C. Vogt, and P. Renaud. 2000. Rapid prototyping of small size objects. *Rapid Prototyping Journal* 6: 259–266.
Gong, H., B. P. Bickham, A. T. Woolley, and G. P. Nordin. 2017. Custom 3D printer and resin for 18 μm× 20 μm microfluidic flow channels. *Lab on a Chip* 17 (17): 2899–2909.
Hay, J. C., A. Bolshakov, and G. M. Pharr. 1999. A critical examination of the fundamental relations used in the analysis of nanoindentation data. *Journal of Materials Research* 14 (6): 2296–2305.
Jacobs, P. F. 1992. Fundamentals of stereolithography. *International Solid Freeform Fabrication Symposium* 196–211.
Manias, E., J. Chen, N. Fang, and X. Zhang. 2001. Polymeric micromechanical components with tunable stiffness. *Applied Physics Letters* 79(11): 1700–1702.
Mirasoli, M., M. Guardigli, E. Michelini, and A. Roda. 2014. Recent advancements in chemical luminescence-based lab-on-chip and microfluidic platforms for bioanalysis. *Journal of Pharmaceutical and Biomedical Analysis* 87: 36–52.
Neumeister, Andre, Roland Himmelhuber, Christian Materlik, Thorsten Temme, Florian Pape, Hans-Heinrich Gatzen, and Andreas Ostendorf. 2008. Properties of three-dimensional precision objects fabricated by using laser based micro stereo lithography. *JLMN-Journal of Laser Micro/Nanoengineering* 3 (2): 67–72.
Prasanna, Raghavendra K., Krishna Prasad, N. Balashanmugam, and G. C. Mohan Kumar. 2013. Surface and dimensional integrity studies of micro polymer components fabricated by micro stereolithography. *3rd International Engineering Symposium (IES 2013)*, Kumamoto University, Kumamoto, Japan.
Yu, H., B. Li, and X. Zhang. 2006. Flexible fabrication of three-dimensional multi-layered microstructures using a scanning laser system. *Sensors and Actuators A: Physical* 125 (2): 553–564.
Zhang, X., X. N. Jiang, and C. Sun. 1999. Micro-stereolithography of polymeric and ceramic microstructures. *Sensors and Actuators A: Physical* 77 (2): 149–156.

# 16 An Insight into Computer Integrated Manufacturing

*Şenol Bayraktar*

## CONTENTS

## 16.1 INTRODUCTION

The computer integrated manufacturing (CIM) term refers to the integration of computers into all parts of the business, from production planning to the design, manufacturing and quality assurance of a product. These parts generally represent symbols, paths of materials, different parts and subassemblies (Kochan 2012). Most of the recent research work on CIM is based on its potential and technological developments. It is clear that computer-controlled production systems cannot expect reasonable economic returns from these without the means to effectively plan and manage devices (Harmonosky and Robohn 1991). The complete implementation of CIM implies the multi-flow of information via an automated process in a business organisation, from an order entry to every stage of the value chain to the delivery of finished parts (Figure 16.1) (Johansen et al. 1995).

According to the CIM structure of a manufacturing business, it is necessary to adopt the stages such as determining production priorities, existing production restrictions in the work area, checking of different processing lines, synchronising of the machines in the production line and performing product quality control,

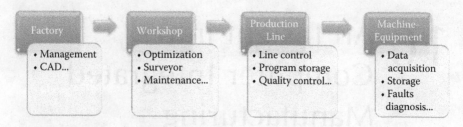

**FIGURE 16.1**   Functional Hierarchy of CIM.

performing final cheques in the machine-equipment area and diagnosing fault (Figure 16.1) (Sandoval 2016). Although the aim of CIM is to promote the development of production systems, it is quite difficult to achieve harmony among different areas and develop a collaborative culture in this field (Brandimarte and Cantamessa 1995). For this reason, it is believed that effective timing and control are the key factors behind the success of CIM systems. Simultaneously, the dynamic control and timing of CIM environments is a concern for manufacturing and academic communities today (Harmonosky and Robohn 1991). CIM systems have been proposed and analysed for many different programming and control methods, such as simulation and artificial intelligence techniques (Harmonosky and Robohn 1991). The advantage of computerised production requires the use of developed and structured CIM philosophy and systems. The CIM philosophy helps production systems integrate computerised systems and their applications in production environments. However, in terms of control, there is a need for resources such as high-tech, expensive equipment and connectivity that are compatible with the latest technology available in the industry (Delaram and Valilai 2018). In order to optimise these resources, they must be part of a CIM system that is able to handle the process and the process flow.

CIM systems generally consist of five functional subsystems. These are CAD – computer-aided design, CAM – computer-aided manufacturing, CAPP – computer-aided process planning, CAQ – computer-aided quality and flexible production systems (FMS – flexible manufacturing systems) (Figure 16.2).

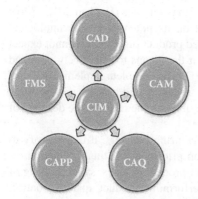

**FIGURE 16.2**   Sub-Support Systems of CIM.

## 16.2 SUB-SUPPORT SYSTEMS OF CIM

### 16.2.1 COMPUTER-AIDED DESIGN AND MANUFACTURING (CAD/CAM)

CIM is defined as the integration of the entire manufacturing organisation that improves organisational and employee productivity with integrated systems and data transmissions. As a result of the development of factory mechanisation, mass production has been accomplished in order to meet the demand of consumers for products. Transfer lines and fixed automation systems have been established for mass production. As a result of these developments, programmable automation systems have emerged (Demirel and Karaağaç 2014). CIM programs that include factory communication hardware and software for planning and control, CAD-CAM and quality and inspection and data management are recognised as a strategic information system that is usually created by technical groups.

The rapid development of computer technologies, networking and communication are important tools for CIM. The CAD/CAM method provides superior and high-quality machining technology and is exceedingly important for customers' prototype production. Moreover, it reduces both product design time and cost. It facilitates indirect production by creating real parts with direct digital input. The CAD/CAM system aims to create an integrated platform used to produce high-quality products with minimum delivery time, in terms of design and automatic machining (Ružarovský 2014). In regard to engineering information flow, the system is integrated with CAD system CAM. The CAD and CAM processes for machining consist of process planning and NC programming, respectively. Developments in CAD applications and CAM-based systems have created the concept of CIM and have been deemed advanced manufacturing technologies (AMTs) (Matsushima et al. 1982). The data obtained from the CAD drawing includes the drawing of the product to be produced by manufacturing methods, such as CNC turning, milling, laser, wire erosion and abrasive water jets. These data are classified as geometric and non-geometric data. CAD provides information not only to assist the designer in the design process, but also to aid in production planning activities (Jabal et al. 2009). The ability to obtain finished products in the operating environment using the CAD/CAM system essentially consists of the steps found in Figure 16.3. In the first stage, the CAD model is designed in any solid model software. In the second step, CAD data with different extensions such as .Step, .Sat, .Iges are obtained according to the requirement. In the third step, NC codes are generated by using tool path generation algorithms according to CAD data in CAM software. In the final stage, visual testing and CMM (coordinate measuring machine) measuring control are done (Figure 16.3) (Sivakumar and Dhanalakshmi 2013).

### 16.2.2 COMPUTER-AIDED PROCESS PLANNING (CAPP)

After the First Industrial Revolution, there have been fundamental changes in production, from water and steam-powered machines to electrical and automatic machines. These changes have led to the development of industrial technologies to thrive on shorter product development time and the efficient use of resources (Qin et al. 2016). Accordingly, this is the point of activation for CIM. CIM is a future growth

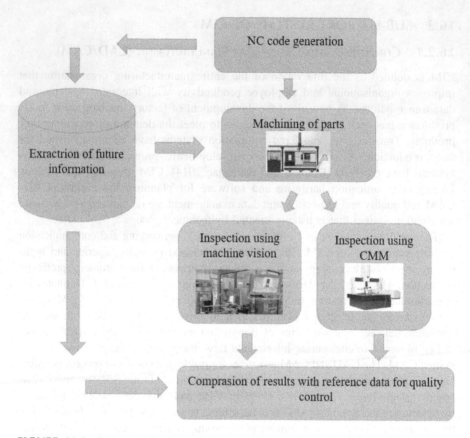

**FIGURE 16.3** Integrated Production Scheme (Reprinted from *International Journal of Computer Integrated Manufacturing,* 26, Sivakumar, S. and Dhanalakshmi, V An Approach Towards the Integration of CAD/CAM/CAI through STEP File Using Feature Extraction for Cylindrical parts, 561–570, with permission of Taylor and Francis).

factor that seeks to achieve sustainable development through the management and improvement of existing production factors such as efficiency, quality, delivery and flexibility (Kang et al. 2016). CIM plays an important role in ensuring efficient management of planning flow and control of materials. CAPP is among the CIM factors that include the most appropriate manufacturing and assembly processes and determine the order in producing a part according to the properties specified in the product design documentation (Groover 2016). Creating a process plan using a computer is referred to as CAPP. The process planning links the design and actual production of any part in a workplace (Basinger et al. 2018). Furthermore, the use of CAPP systems reduces the time required to plan parts manufacturing (Yusof and Latif 2014). It is important to shorten production times in order to achieve high efficiency in the manufacturing industry. Therefore, a computer-aided process planning (CAPP) system has been developed for businesses (Koremura et al. 2017). CAPP and production planning-control (PPC) integration is carried out to eliminate the concern for the product life cycle (El Maraghy 1993). CAPP systems are widely used for individual processes and recently, the development of CAPP systems has been re-directed to mixed operations that are feature-based or feature-less systems (Basinger et al. 2018). Furthermore, some basic features are required in a CAPP system. The first is the connection between CAD/CAM to produce products, and the processes are expandable, adaptable and customisable. The second is the ability to obtain effective information in order to check the integrity and consistency of this information. Third, users can include the information obtained in some parts of the decision-making process, the ability to perform heuristic scanning when necessary, support the cap-abilities of the system, produce outputs by facilitating inputs and showing the results graphically, and, finally, becoming effective by showing results (Xu et al. 2011).

Although designed to operate as an independent system, the CAPP is integrated into the CAD/CAM system (Figure 16.4). Design and manufacturing engineers develop and use CAD/CAM systems using the same computer database in processes where parts can be produced from the online designs, hence eliminating any paperwork. Thus, integration is important where the same database and computers are prioritised. The process plan can ensure without human intervention when the computer is con-nected to the targeted CNC machine. This makes it possible to use CAPP systems and further enables these to be used more efficiently (Architecture Technology Corp. 1991). The integration of CAD/CAM with CAPP shortens the production cycle and increases productivity by enabling manufacturing companies to benefit greatly (Xu et al. 2015). CIM is recognised as an effective platform for increasing production competitiveness. Ultimately, CAPP is the key to reaching CIM (El Maraghy 1993). Process planning has traditionally been regarded as production preparation that provides production methods and operating instructions. When CAPP was introduced, it emerged as the vital link between CAD and CAM. Process planning starts with the design of the product for error-free production. Next, CAD data are defined in the technological environment of the product designed in the specified dimensions. It is ensured that the necessary documents are obtained by processing the data obtained. In order to carry out these steps, complete support is necessary from the CAPP system (Figure 16.5) (Schachter-Radig and Wermser 1990).

### 16.2.3 Flexible Manufacturing Systems (FMS)

FMS emerges as a part of CIM that ensures the integration of material and op-erational capacity in order to eliminate problems arising from lack of material and information, as well as to increase capacity utilisation rate (Ibraimi et al. 2016). FMS contributes to CIM integration with programmable automation through the

**FIGURE 16.4**   CAPP-CAD-CAM Relationship.

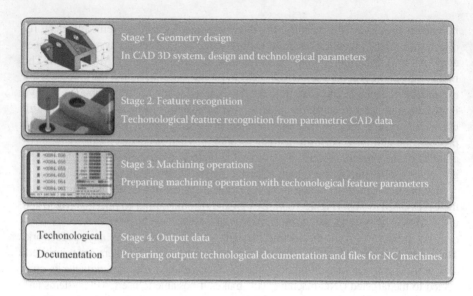

**FIGURE 16.5** The Four Main Stages of Planning the Production Processes in a CAPP System (Reprinted from *Computer Integrated Manufacturing,* Schachter-Radig, M.J. and Wermser, D, *Quality Assurance as a Dynamical Production Process Guide-Control Elements Supported by Dedicated Knowledge-Based Systems,* 233–244, 1990, with permission of Springer).

machines used. It is also designed to provide integration among parts for the partial production of similar parts (Sandoval 2016). Computer control systems are used to coordinate the operation and momentary status of the processing stations, material transportation systems in FMS (Shivanand et al. 2006). Therefore, integrated computer-controlled automatic material transportation devices and numerically controlled machine tools that can process the desired volume or volumes of various part types are the main components of FMS (Figures 16.6 and 16.7). This system has been developed to improve the efficiency of well-balanced transfer lines for

**FIGURE 16.6** Main Structure of FMS (Reprinted from Mahmood et al. (2017), with permission of Elsevier).

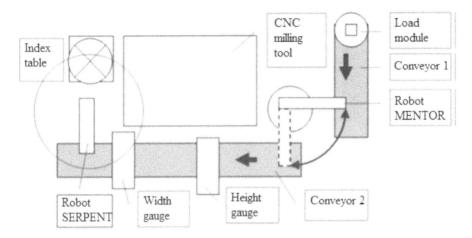

**FIGURE 16.7**   Flexible Production System (Reprinted from Procedia CIRP, 63, Mahmood, K., Karaulova, T., Otto, T. and Shevtshenko, E, Performance Analysis of a Flexible Manufacturing System (FMS), 424–429, 2017, with permission of Elsevier).

simultaneous processing of multiple part types (Browne et al. 1984). Flexible production systems are generally preferred for the production of multi-product, technological, biological, pharmaceutical or chemical products in expandable plants (Scannon et al. 2017).

Thanks to people, who constitute one of the main elements of FMS, production systems are improving. People play an important role in the functioning, basically, the success, of the production systems and therefore the competitiveness of the businesses. Because of the quest for productivity and profitability, businesses have realised that people are critical to decision-making, success and competitiveness in production systems (El Maraghy 2005). It is necessary to integrate human-machine systems with flexible production systems by means of automation (Figure 16.8).

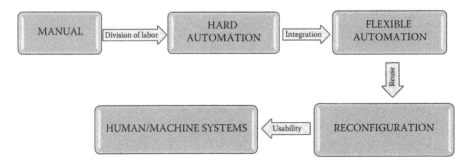

**FIGURE 16.8**   The Role of Human and Automation in the Development of Manufacturing Systems (Reprinted from *International Journal of Flexible Manufacturing Systems*, 17, El Maraghy, H.A, Flexible and reconfigurable manufacturing systems paradigms, 261–276, 2005, with permission of Springer).

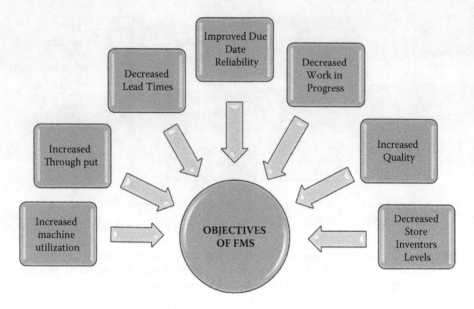

**FIGURE 16.9**  Main Objectives of Flexible Production Systems.

The main objective in FMS is group technology based on the principle of material flow analysis. It also consists of several factors such as reducing delivery times, minimising manpower, increasing the efficiency of machinery, ensuring better use of production equipment, improving technical performance, reducing part cost and improving quality (Figure 16.9) (Kaushal et al. 2016).

FMS requires the use of computerised numerical controls and other automation techniques to provide flexibility in operations, machinery, products, quality and processes (Jamali et al. 2015). These techniques are important in terms of increasing product quality and performance, ensuring continuity in production, using optimum process times and performing useful modifications in terms of production volume or processes (Figure 16.10) (Singh et al. 2018).

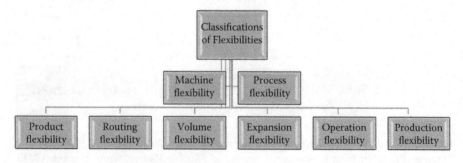

**FIGURE 16.10**  Classification of Flexible Production Systems.

## 16.2.4 COMPUTER-AIDED QUALITY (CAQ)

Recipients of manufactured products apply to quality assurance programs in order to ensure that products are reproduced at optimum quality. In addition, product suppliers use different electronic verification and verification software systems to be able to control the quality of the manufacturing or manufacturing process (Boyer 2008). Product quality output is generally planned to be high. To achieve the desired quality, first, the starting point or boundary conditions are determined in terms of orientation parameters. Accordingly, Wolf et al. (2000) conducted an experimental study on the CAQ control approach based on 3D coordinate metrology. Initially, 3D data were obtained by using an optical 3D measurement method with reference to any product manufactured before. Numerical algorithms were used for a minimum distance function during data recording. With the obtained approach, it has been discovered that quality control is provided for all kinds of rapid prototyping or NC-produced objects, and this approach is particularly suitable for integration into CAQ processes (Wolf et al. 2000). Consequently, CAQ approaches can be used to perform processes, such as the monitoring of tasks required for mass production of mechanical parts, tolerance analysis and change of worn parts. In particular, CAQ approaches can be utilised for the rapid production of components with critical dimensions in a high volume production (Gallagher and Kurfess 1998). In addition, technological synchronisation in the chain of supporting activities during the checking and validation processes can be achieved with different CAQ software (Boyer 2008). In order to make the product development process more efficient, advanced CAQ approaches play an active role in applying this to an independent system and operate on the most powerful hardware available today (Wolf et al. 2000). Quality assurance requires not only that the many products are to be produced and the tools and data be applied, but rather, the semantic integration of production as well (Figure 16.11). Therefore, assuming that CAQ is a component of CIM, their integration needs to be extremely well executed in order to be able to carry out processes in CIM more efficiently. Accordingly, the success of CAQ results from integration (Schachter-Radig and Wermser 1990).

## 16.2.5 MAINTENANCE

Maintenance can generally be defined as a set of all of the technical and administrative activities, including actions and ideas, that are aimed at restoring the system from a state where it cannot perform a necessary function (Ostadi 2018). Maintenance consists of repair and overhaul. Machines and different processes in flexible production systems have the ability to achieve a specific purpose. Failure of the machine is usually due to operations performed on the machine. In order to ensure uninterrupted production and to increase efficiency, regular maintenance of the system is paramount. Maintenance is essential for the service life of machines and CIM. It is also important to examine the maintenance operations in terms of cost. A mechanical system that is considered to be serviced, as well as cost analysis between the estimated service life of this system and the new system, are some of the factors affecting production efficiency. If it is necessary and if a better-quality product will be obtained in terms of tolerance and dimensional integrity with the

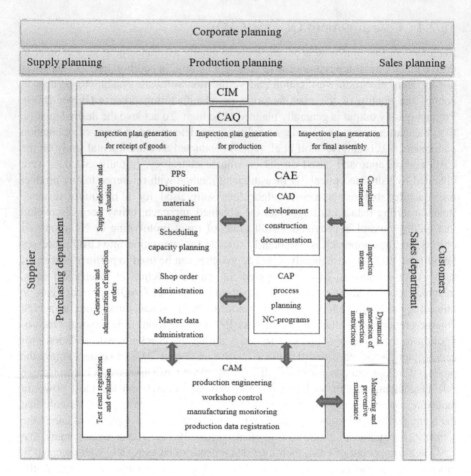

**FIGURE 16.11** CAQ as an Integrating Factor in CIM (Reprinted from *Computer Integrated Manufacturing, First Edition,* Schachter-Radig, M.J. and Wermser, D, *Quality Assurance as a Dynamical Production Process Guide-Control Elements Supported by Dedicated Knowledge-Based Systems,* 233–244, 1990, with permission of Springer).

new mechanical system or machine, then it is useful to choose this. However, in order to successfully determine this, criteria such as the number of products, quality and tolerance interval should be taken into consideration. Maintenance procedures in businesses differ according to their purposes (Figure 16.12).

The most common type of maintenance used in the manufacturing industry is the corrective maintenance policy. Corrective maintenance policy aims to repair and restore the old order of the system with a simple operation (Ding and Kamaruddin 2015). Preventive maintenance is preferred in performing the necessary processes by detecting possible faults in the system. Improvement maintenance is performed to enhance the workings of the current system. It is of utmost importance to determine the ideal maintenance time for CIM in businesses. The experience of the operators and different optimisation techniques are used to determine this. Identifying the ideal

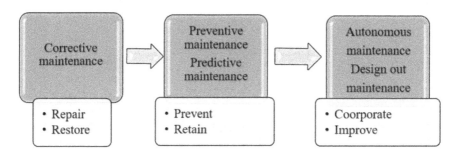

**FIGURE 16.12** Classification of Maintenance Policies (Reprinted from *International Journal of Advanced Manufacturing Technology, 76,* Ding, S.H. and Kamaruddin, Maintenance Policy Optimisation-Literature Review and Directions, 1263–1283, 2015, with permission of Springer).

models for maintenance operations and applying them in line with these models is an important step in the efficiency of maintenance operations (Scheer 2012). In the resolution of models, graphical, simulation, artificial intelligence and multi-criteria decision making (MCDM) categories are advisable under three main headings such as certainty, risk and uncertainty (Ding and Kamaruddin 2015). Thus, the use of ideal techniques for maintenance procedures, which is one of the main components of CIM, is suitable in terms of time and cost factors.

## 16.3 ADVANCEMENTS AND LATEST TECHNOLOGY IN CIM

Nowadays, two features stand out among diverse qualities that characterise modern production technology, such as simple, fast and virtual response systems and in relation to the next generation of manufacturing. These are integrated production and smart manufacturing. Figure 16.13 presents the new inclination of manufacturing systems based on big data analytics and production environment (Chen 2017).

The development of integrated and smart manufacturing technology ($i^2M$) is determined not only by market demand, but also by technological advancements. For the success of integrated and intelligent manufacturing, ten key technologies are crucial (see Figure 16.14). 3D printing or additive manufacturing, robotic automation, advanced materials, virtual or augmented reality, the industrial internet and cyber-physical systems (CPSs) are important. The four other key elements are big data analytics, cloud computing, applications and mobile devices (Chen 2017).

The concept of manufacturing systems began to become defined through advances in digital computing capability in the 1960s. At this point, the need for an integration process has started to emerge in terms of manufacturing. Accordingly, the devices and machines used in manufacturing have not been independently considered. In order to achieve high production efficiencies, these devices have been accepted as part of the system, and effective coordination can be achieved among them. The internet of things (IoT), cyber-physical systems (CPSs) and cloud computing (CC) technologies play an important role in making this integration broader and more explicit. Thus, manufacturing systems are not limited to physical components such as

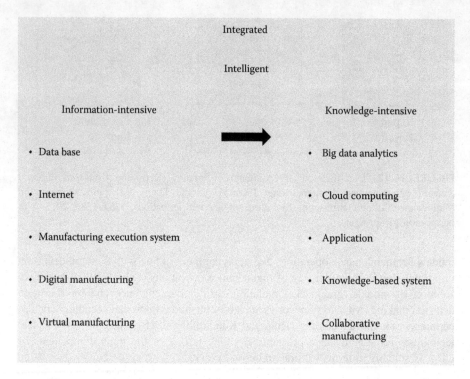

**FIGURE 16.13**   The New Inclinations in Manufacturing Systems (Chen 2017).

materials, machinery and equipment, but rather, a wide variety of information and data can be processed simultaneously (Chen 2017; Kagermann et al. 2013).

### 16.3.1 Internet of Things (IoT)

In recent years, IoT has been a major topic of interest for researchers, scholars and engineers (Atzori et al. 2010; Tsai et al. 2014; Sanin et al. 2019). In IoT, the main

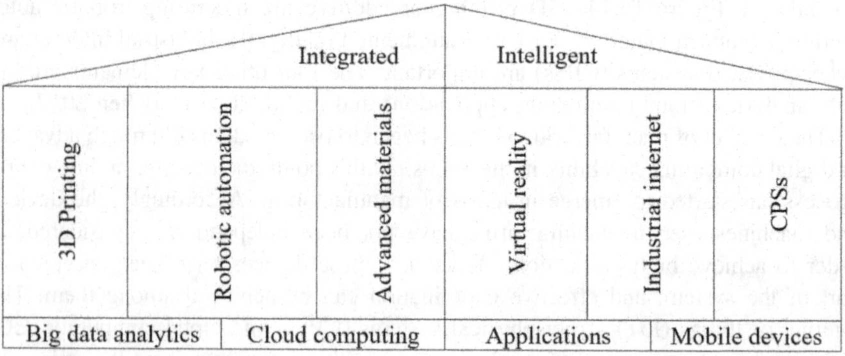

**FIGURE 16.14**   Ten Major Technologies for Integrated and Intelligent Manufacturing (Chen 2017).

objective is to facilitate human life by creating efficient information exchange, ensuring interconnectivity and providing effective communication (Sanin et al. 2019). In particular, this provides a new method for intelligent detection and connection of virtually anything. Generally, an intelligent system is developed by IoT. Thus, IoT and CC have been examined and applied in many areas of domestic, industrial and commercial relevance. IoT is widely used in smart homes, buildings, cities, inventory and product management, as well as in social security and healthcare (Jeschke et al. 2017; Miorandi et al. 2012). In particular, it fundamentally changes the process and management of practical production and supply chain to intelligent production. Depending on the integration of IoT into the manufacturing industry, resources and feasibility in manufacturing are achieved through connection, communication, computing, and control (4C) (Figure 16.15) (Jeschke et al. 2017).

Despite many advances in technology, manufacturers still face some difficulties related to information capture, integration, synchronisation and distribution of IoT technologies; an example of which is RFID (radio-frequency identification). By taking into consideration the advantages of IoT and the optimal decision requirements of the real-time data-driven production application system, the internet of manufacturing things (IoMT) can be captured and expanded by bolstering IoT into the production area. Any changes can be monitored and integrated with enterprise information systems (EISs). The information gap between corporate management information systems and workstations, trolleys and similar representative executive units can be eliminated with IoMT. Thus, real-time and information-oriented optimum control and decision-making can be facilitated during manufacturing execution (Zhang et al. 2015).

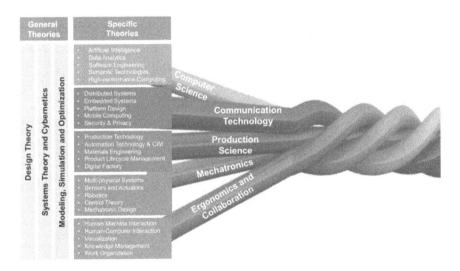

**FIGURE 16.15**   Different Applications of IoT in Manufacturing (Reprinted from Industrial Internet of Things and Cyber Manufacturing Systems, First Edition, Jeschke, S., Brecher, C., Meisen, T., Özdemir, D. and Eschert, T 2017, with permission of Springer).

## 16.3.2 CYBER-PHYSICAL SYSTEMS (CPSs)

CPSs consist of software and hardware components connected via a common network in order to control physical processes in various domains, such as smart grids, autonomous car systems, process control systems, robotic systems and autopilot avionics. The architecture of CPS remains exceedingly complex, because it must be capable enough to monitor physical processes, make decentralised decisions and trigger actions. Moreover, the devices based on this system must communicate in real-time, collaborate with one another and interact with people (Khaitan and McCalley 2015; Lins et al. 2020). As a result of the demand for new products in the globalising world, a new technique should be applied to manufacturing systems in order to guarantee talent, sensitivity and product quality. Therefore, CPSs are emerging as an alternative to overcome the challenges of manufacturing systems. It was named cyber-physical production systems (CPPSs) by adapting manufacturing systems to CPSs systems (Lu et al. 2020). CPSs acclimate quickly and effectively to new factory manufacturing systems and new products. The purpose of adaptive CPSs is modularity and reusability (Katzke et al. 2004; Wang et al. 2015). CPS-based smart manufacturing, with the intervention of other modern technologies, can make value addition to the product more prominent (Wang et al. 2015; Yao et al. 2019a). A CPPS consists of eight parts, namely input, relation, CPS, IoS, IoT, IoCK, factory and output (Figure 16.16). Input refers to the set of input items, including customer requirements, materials, energy, capital and labour. IoS (i.e. internet of services) makes use of cloud computing and web technologies to perform connection, communication and interaction. IoCK (internet of content and knowledge) refers to data and

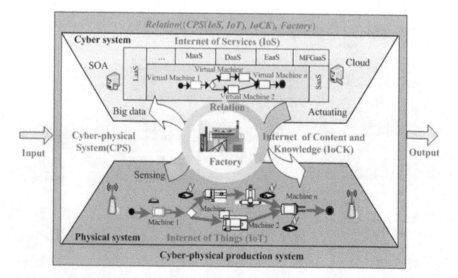

**FIGURE 16.16** Structure of CPS-Based Manufacturing (Reprinted from *Journal of Intelligent Manufacturing*, 30, Yao, X., Zhou, J., Lin, Y., Li, Y., Yu, H. and Liu, Y., Smart Manufacturing based on Cyber-Physical Systems and Beyond, 2805–2817, 2019b, with permission of Springer).

information. Subsequently, system interaction, joint action and collaboration among IoT, IoS IoCK and CPS are represented by relation. Factory represents the location where the raw material is converted to finished product with the support of IoT, IoS, IoCK and CPS according to the relationship in the production processes, while output represents the content set of output items, such as products and/or services as well as solutions (Yao et al. 2019b).

A CPS architecture was proposed for a type of human-robot collaboration that was considered industrially safe (Figure 16.17) (Nikolakis et al. 2019). They stated that they focused on the response time and trigger action signal required to detect the human in the factory working environment. As a result, they discovered that if both detection and response time are short, then accidents can be avoided and the safety of the operators can be successfully ensured (Nikolakis et al. 2019).

In another study, Lins et al. (2020) confirmed the monitoring of tool wear by performing experimental work on the CNC (i.e. computer numerical control) machine for CPSs (Figure 16.18). The main purpose of CPSs is mainly to raise requirements, verify integration, and improve decentralised architecture. In line with this, a four-stage methodology has been proposed to demonstrate the competence, robustness and capabilities that it can provide to a production system. The first is the need analysis for the proper development of the system. The second is to make Petri net-based dynamic modelling in order to ensure integration between CNC and computer monitoring systems. The third is the integration of LUPA through decentralised architecture, which plays an important role in monitoring tool wear after requirement and verification results. Lins et al. 2020 recommended the use of the system for tool wear when monitoring a CNC drilling operation.

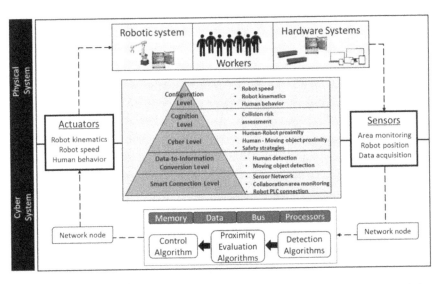

**FIGURE 16.17** Hierarchical Diagram of the Applied CPS Approach (Reprinted from *Robotics and Computer-Integrated Manufacturing, 56,* Nikolakis, N., Maratos, V. and Makris, S., A Cyber-Physical system (CPS) Approach for Safe Human-Robot Collaboration in a Shared Workplace, 233–243, 2019, with permission of Elsevier).

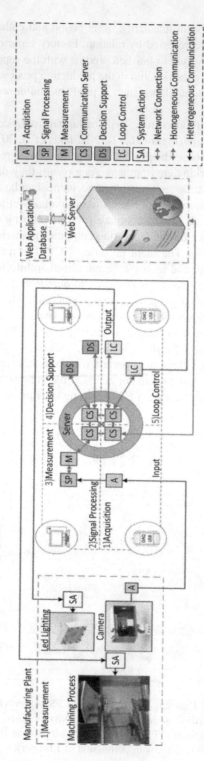

**FIGURE 16.18** Decentralised Structure for CPS Monitoring System (Reprinted from *Robotics and Computer-Integrated Manufacturing, 61*, Lins, R.G., Araujo, P.R.M. and Corazzim, M., In-Process Machine Vision Monitoring of Tool Wear for Cyber-Physical Production Systems, 101859, 2020, with permission of Elsevier).

### 16.3.3 CLOUD COMPUTING (CC)

CC is an internet-based computing technology that is based on optional access of shared resources and used as 'pay-as-you-go' (Wei and Blake 2010). In this paradigm, users receive high-quality service at a low cost. It draws great attention from academia and industry because it is socially and economically relevant. For example, many companies, such as Amazon, Google, Salesforce, IBM and Microsoft, are developing cloud computing platforms to serve users (Zhang et al. 2010; Qi and Tao 2019). At the same time, the structure, in combination with CC manufacturing, contributed to the emergence of the cloud-based manufacturing (CMfg) model. All production resources and capabilities are virtualized in CMfg. Management, allocation and services are encapsulated to be requested via CC (Qi and Tao 2019; Tao et al. 2011). Network, CC, service computing (SC) and production technologies can be used to transform resources and capabilities. Moreover, this system can be managed in a smart and unified way to enable full sharing and circulation of production resources and production capabilities (Zhang et al. 2014, 2017). CC transforms its traditional production models into a computing and service-oriented production model. The structure of the CC-based manufacturing system can be summarised as given in Figure 16.19 (Brant and Sundaram 2015). The design of this manufacturing system consists of the physical manufacturing system and the cloud environment. The substrate represents users who are participating in manufacturing resources and capabilities and production activities, including production devices.

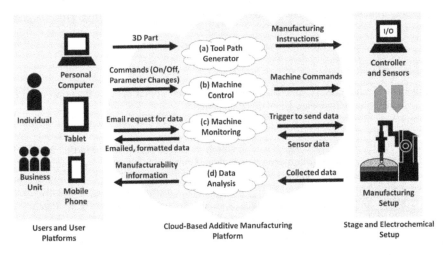

**FIGURE 16.19** Structure of Cloud-Based Manufacturing System (Reprinted from Journal of Manufacturing Processes, *20,* Brant, A. and Sundaram, M.M., A Novel System for Cloud-Based Micro Additive Manufacturing of Metal Structures, 478–484, 2015, with permission of Elsevier).

## 16.4 CONCLUDING REMARKS

CIM is one of the most important methods of system integration in manufacturing, as well as success in decision-making. All inputs, outputs and all factors between them can be controlled by using a high-speed computer for CIM. Thus, integration among the departments can be carried out more accurately by measuring, controlling, reporting and communicating among CNC, CAD/CAM, FMS, CAPP and CAQ processes which are included in the components of integrated manufacturing. In addition, the loss of time and personnel errors between the departments can be minimised by operating the control mechanism in the businesses, and it makes a significant contribution to increasing productivity in manufacturing. With CIM, controlling capacity in businesses and instituting necessary changes in processes in any department can also be simply accomplished. Accordingly, the efficient use of all resources and the sustainability of the competitiveness of the businesses are ensured by the use of CIM, particularly in enterprises producing large volumes. Different requirements must be satisfied in order to realise the factory visions of the future. The fact that the human being is the most flexible asset in the production systems forces it to constantly follow the changing technology. Therefore, people play an important role in criteria such as using, updating and distributing information in manufacturing systems. Moreover, the capabilities of the components that constitute the manufacturing systems and the coordination among them are closely related to the use of human information on the manufacturing system. Different approaches, such as ensuring integration between human and manufacturing systems, filling the gap between products, processes and manufacturing systems, synchronisation in an efficient and information-oriented environment, computer-based CIM, IoT, CPS and CC, are advisable. In this chapter, the role of these approaches on integrated manufacturing systems and their effects on manufacturing efficiency are deliberated.

## REFERENCES

Architecture Technology Corp. (Auth.). 1991. Computer-Aided Process Planning (CAPP). Minneapolis, Minnesota, USA: Elsevier Science.

Atzori, L., A. Iera, and G. Morabito. 2010. The internet of things: A survey. *Computer Networks* 54: 2787–2805.

Basinger, K. L., C. B. Keough, C. E. Webster, R. A. Wysk, T. M. Martin, and O. L. Harrysson. 2018. Development of a modular computer-aided process planning (CAPP) system for additive-subtractive hybrid manufacturing of pockets, holes, and flat surfaces. *International Journal of Advanced Manufacturing Technology* 96: 2407–2420.

Boyer, M. J. 2008. Computer aided quality assurance software system. U.S. Patent No: 7,327,869.

Brandimarte, P., and M. Cantamessa. 1995. Methodologies for designing CIM systems: A critique. *Computers in Industry* 25: 281–293.

Brant, A., and M. M. Sundaram. 2015. A novel system for cloud-based micro additive manufacturing of metal structures. *Journal of Manufacturing Processes* 20: 478–484.

Browne, J., D. Dubois, K. Rathmill, S. P. Sethi, and K. E. Stecke. 1984. Classification of flexible manufacturing systems. *The FMS Magazine* 2: 114–117.

Chen, Y. 2017. Integrated and intelligent manufacturing: Perspectives and enablers. *Engineering* 3: 588–595.

Delaram, J., and O. F. Valilai. 2018. An architectural view to computer integrated manufacturing systems based on axiomatic design theory. *Computers in Industry* 100: 96–114.

Demirel, M. Y., and İ. Karaağaç. 2014. An overview of computer aided manufacturing process. *Engineer and The Machinery Magazine* 55: 51–61.

Ding, S. H., and S. Kamaruddin. 2015. Maintenance policy optimization-literature review and directions. *International Journal of Advanced Manufacturing Technology* 76: 1263–1283.

El Maraghy, H. A. 1993. Evolution and future perspectives of CAPP. *CIRP Annals* 42: 739–751.

El Maraghy, H. A. 2005. Flexible and reconfigurable manufacturing systems paradigms. *International Journal of Flexible Manufacturing Systems* 17: 261–276.

Gallagher, C. T., and T. R. Kurfess. 1998. Design and implementation of a system for rapid inspection of critical dimensions in high volume production. *Mechatronics* 8: 413–425.

Groover, M. P. 2016. *Automation, Production Systems, and Computer-integrated Manufacturing.* India: Pearson Education.

Harmonosky, C. M., and S. Robohn. 1991. Real-time scheduling in computer integrated manufacturing: a review of recent research. *International Journal of Computer Integrated Manufacturing* 4: 331–340.

Ibraimi, S., A. Bexheti, R. Zuferi, G. Rexhepi, and V. Ramadani. 2016. Enhancing flexible manufacturing competence. *The Eurasia Proceedings of Educational and Social Sciences* 5: 378–384.

Jabal, M. F., M. S. Rahim, N. Z. Othman, and D. Daman. 2009. Computer-aided design data extraction approach to identify product information. *Journal of Computer Science* 5: 624.

Jamali, G., K. Farrokhnejad, and M. Mohammadi 2015. Decision making on analyzing advanced manufacturing systems dimensions: SWARA and COPRAS G integration (case study: automotive industry). *Buletin Teknologi Tanaman* 12: 266–274.

Jeschke, S., C. Brecher, T. Meisen, D. Özdemir, and T. Eschert. 2017. *Industrial Internet of Things and Cyber Manufacturing Systems.* Cham: Springer.

Johansen, J., U. S. Karmarkar, D. Nanda, and A. Seidmann. 1995. Computer integrated manufacturing: empirical implications for industrial information systems. *Journal of Management Information Systems* 12: 59–82.

Kagermann, H., J. Helbig, A. Hellinger, and W. Wahlster. 2013. Recommendations for implementing the strategic initiative INDUSTRIE 4.0: Securing the future of German manufacturing industry. *Final report of the Industrie 4.0 Working Group.* Forschungsunion.

Kang, H. S., J. Y. Lee, S. Choi, H. Kim, J. H. Park, J. Y. Son, and N. S. Do. 2016. Smart manufacturing: Past research, present findings, and future directions. *International Journal of Precision Engineering and Manufacturing-Green Technology* 3: 111–128.

Katzke, U., K. Fischer, and B. Vogel-Heuser. 2004. Development and evaluation of a model for modular automation in plant manufacturing. In 10th International Conference Information Systems Analysis and Synthesis (CITSA), pp. 15–20.

Kaushal, A., A. Vardhan, and R. S. Rajput. 2016. Flexible manufacturing system a modern approach to manufacturing technology. *International Refereed Journal of Engineering and Science* 5: 16–23.

Khaitan, S. K., and J. D. McCalley. 2015. Design techniques and applications systems: a survey. *IEEE Systems Journal* 9: 350–365.

Kochan, D. 2012. *CAM: Developments in Computer-Integrated Manufacturing.* Berlin: Springer.

Koremura, K., Y. Inoue, and K. Nakamoto. 2017. Machining process evaluation indices for developing a computer aided process planning system. *International Journal of Automation Technology* 11: 242–250.

Lins, R. G., P. R. M. Araujo, and M. Corazzim. 2020. In-process machine vision monitoring of tool wear for cyber-physical production systems. *Robotics and Computer-Integrated Manufacturing* 61: 101859.

Lu, Y., C. Liu, I. Kevin, K. Wang, H. Huang, and X. Xu 2020. Digital twin-driven smart manufacturing: connotation, reference model, applications and research issues. *Robotics and Computer-Integrated Manufacturing* 61: 101837.

Mahmood, K., T. Karaulova, T. Otto, and E. Shevtshenko. 2017. Performance analysis of a flexible manufacturing system (FMS). *Procedia CIRP* 63: 424–429.

Matsushima, K., N. Okada, and T. Sata. 1982. The integration of CAD and CAM by application of artificial-intelligence techniques. *CIRP Annals* 31: 329–332.

Miorandi, D., S. Sicari, F. De Pellegrini, and I. Chlamtac. 2012. Internet of things: Vision, applications and research challenges. *Ad Hoc Networks* 10: 1497–1516.

Nikolakis, N., V. Maratos, and S. Makris. 2019. A cyber physical system (CPS) approach for safe human-robot collaboration in a shared workplace. *Robotics and Computer-Integrated Manufacturing* 56: 233–243.

Ostadi, B. 2018. An optimal preventive maintenance model to enhance availability and reliability of flexible manufacturing systems. *Journal of Industrial Systems Engineering* 11: 47–61.

Qi, Q., and F. Tao. 2019. A smart manufacturing service system based on edge computing, fog computing, and cloud computing. *IEEE Access* 7: 86769–86777.

Qin, J., Y. Liu, and R. Grosvenor. 2016. A categorical framework of manufacturing for industry 4.0 and beyond. *Procedia CIRP* 52: 173–178.

Ružarovský, R. 2014. Direct production from CAD models considering on integration with CIM flexible production system. *Applied Mechanic and Materials* 474: 103–108.

Sandoval, V. 2016. *Computer Integrated Manufacturing (CIM) in Japan*. Netherlands: Elsevier.

Sanin, C., Z. Haoxi, I. Shafiq, M. M. Waris, C. S. Oliveira, and E. Szczerbicki. 2019. Experience based knowledge representation for internet of things and cyber physical systems with case studies. *Future Generation Computer Systems* 92: 604–616.

Scannon, P. J., F. Bernard, A. C. Dadson, and R. S. Tenerowicz. 2017. U.S. Patent Application No. 15/349, 298.

Schachter-Radig, M. J., and D. Wermser. 1990. Quality Assurance as a Dynamical Production Process Guide-Control Elements Supported by Dedicated Knowledge Based Systems, eds. Faria, L., and Van Puymbroeck, W., 233–244. In *Computer Integrated Manufacturing*. London: Springer.

Scheer, A. W. 2012. *CIM Computer Integrated Manufacturing: Towards the Factory of the Future*. Berlin: Springer.

Shivanand, H. K., M. M. Benal, and V. Koti. 2006. *Flexible Manufacturing System*. New Delhi, India: New Age International.

Singh, A., J. Singh, and M. Ali. 2018. Some control strategies in a flexible manufacturing system-A simulation perspective. *International Journal of Applied Engineering Research* 13: 5296–5303.

Sivakumar, S., and V. Dhanalakshmi. 2013. An approach towards the integration of CAD/CAM/CAI through STEP file using feature extraction for cylindrical parts. *International Journal of Computer Integrated Manufacturing* 26: 56–57.

Tao, F., L. Zhang, V. C. Venkatesh, Y. Luo, and Y. Cheng. 2011. Cloud manufacturing: A computing and service-oriented manufacturing model. *Proceedings of the Institution of Mechanical Engineers, Part B: Journal of Engineering Manufacture* 225: 1969–1976.

Tsai, C., C. Lai, M. Chiang, and L. Yang. 2014. Data mining for internet of things: A survey. *IEEE Communications Surveys & Tutorials* 16: 77–97.

Wang, L., M. Törngren, and M. Onori. 2015. Current status and advancement of cyber-physical systems in manufacturing. *Journal of Manufacturing Systems* 37: 517–527.

Wei, Y., and M. B. Blake. 2010. Service-oriented computing and cloud comput-ing: Challenges and opportunities. *IEEE Internet Computing* 14: 72–75.

Wolf, K., D. Roller, and D. Schäfer. 2000. An approach to computer-aided quality control based on 3D coordinate metrology. *Journal of Material Processing Technology* 107: 96–110.

Xu, T., Z. Chen, J. Li, and X. Yan. 2015. Automatic tool path generation from structuralized machining process integrated with CAD/CAPP/CAM system. *International Journal of Advanced Manufacturing Technology* 80: 1097–1111.

Xu, X., L. Wang, and S. T. Newman. 2011. Computer-aided process planning-A critical review of recent developments and future trends. *International Journal of Computer Integrated Manufacturing* 24: 1–31.

Yao, Y., M. Liu, J. Du, and L. Zhou. 2019a. Design of a machine tool control system for function reconfiguration and reuse in network environment. *Robotics and Computer-Integrated Manufacturing* 56: 117–126.

Yao, X., J. Zhou, Y. Lin, Y. Li, H. Yu, and Y. Liu. 2019b. Smart manufacturing based on cyber-physical systems and beyond. *Journal of Intelligent Manufacturing* 30: 2805–2817.

Yusof, Y., and K. Latif. 2014. Survey on computer-aided process planning. *International Journal of Advanced Manufacturing Technology* 75: 77–89.

Zhang, Q., L. Cheng, and R. Boutaba. 2010. Cloud computing: State-of-the-art and research challenges. *Journal of Internet Services and Applications* 1: 7–18.

Zhang, L., Y. L. Luo, F. Tao, B. H. Li, L. Ren, X. Zhang, H. Guo, Y. Cheng, A. Hu, and Y. Liu. 2014. Cloud manufacturing: A new manufacturing paradigm. *Enterprise Information Systems* 8:167–187.

Zhang, Y., G. Zhang, Y. Liu, and D. Hu. 2017. Research on services encapsulation and virtualization access model of machine for cloud manufacturing. *Journal of Intelligent Manufacturing* 28: 1109–1123.

Zhang, Y., G. Zhang, J. Wang, S. Sun, S. Si, and T. Yang. 2015. Real-time information capturing and integration framework of the internet of manufacturing things. *International Journal of Computer Integrated Manufacturing* 28: 811–822.

# Index